N. CAROLINA

DISMAL
SWAMP

PINETOWN
POCOSIN

S. CAROLINA

GEORGIA

Ocmulgee R.

Brier Cr.

Savannah R.

Edisto R.

Santee R.

I'ON SWAMP

BRIER
CREEK

Charleston

EDISTO RIVER
(JACKSONBORO)

Altamaha R.

OCMULGEE
RIVER

Darien

ALTAMAHA RIVER

DUDLEY'S
HAMMOCK

OKEFENOKEE SWAMP

Valdosta

Suwannee R.

FLORIDA

ATLANTIC OCEAN

MEXICO

Lake
Okeechobee

BIG CYPRESS

EVERGLA

Swamps, River Bottoms and Canebrakes

SWAMPS, RIVER BOTTOMS AND CANEBRAKES

By Brooke Meanley

1972

BARRE PUBLISHERS

Barre, Massachusetts

Library of Congress Catalog Number 73-186366
International Standard Book Number 0-8271-7208-7
Designed by Klaus Gemming, New Haven, Connecticut
Printed in the United States of America
by The Meriden Gravure Company, Meriden, Connecticut

Table of Contents

Acknowledgments

I AM INDEBTED to many of my naturalist friends for permission to use their photographs, for editing, identification of plants, and numerous other favors. I thank them all. Those who have been colleagues of mine in the U.S. Fish and Wildlife Service are Peter J. Van Huizen, Paul A. Stewart, Frederick C. Schmid, Luther C. Goldman, Neil Hotchkiss, James D. Stephenson, Francis M. Uhler, John S. Webb, and E. O. (Mel) Mellinger. Others are Marie Mellinger, naturalist with the Georgia Department of State Parks, Ivan R. Tomkins, noted ornithologist of the Georgia Coast country, Francis Harper, pioneer scientist of Okefenokee Swamp explorations, and John W. Taylor, Chesapeake Bay waterfowl artist.

My wife, Anna Gilkeson Meanley, often waded up to her waist in swamp waters infested with cottonmouths, and up to her knees in the guano of huge blackbird roosts in accompanying me and helping in various projects. I am grateful to her for all her assistance.

I wish also to thank Gale Monson, editor of the *Atlantic Naturalist,* and George A. Hall, editor of the *Wilson Bulletin,* for permission to use excerpts from articles of mine that appeared in these journals, the Harvard College Library for permission to use a photograph of John Abbot's painting of the Swainson's warbler, the Charleston (S.C.) Museum for the use of a photograph of John Bachman, and the National Park Service for the use of other photographs.

Photographs are by the author unless otherwise indicated.

The raccoon is the best known mammal of the swamps and river bottoms.

Introduction

To many persons, a Southern swamp is a forbidding place where poisonous snakes hang from vines, quicksand and alligators are encountered, and one may easily get lost. Much of this is fantasy. Poisonous snakes of the South generally are not tree climbers, and most of them are sluggish in their movements compared with non-poisonous snakes, some of which do climb. One year I spent fifteen days in the Great Dismal Swamp in May and June and saw only six snakes, all non-poisonous. I have never seen quicksand in a swamp, although it may exist in some (however, I have encountered peat bogs in the Dismal Swamp and Okefenokee that seemed bottomless), and today most of the alligators live in wildlife refuges. People who stay on roads and trails rarely get lost, although in vast tracts of swamp or bottomland one should use a compass or guide when leaving the trail. By and large, a Southern swamp is one of the safest places in the world!

The one situation that I was most interested in avoiding, however, was a confrontation with moonshiners. Some of the time I had a gun and binoculars, and with such paraphernalia would be suspect in a moonshiner's territory. On one occasion when following a path in Boeuf River Swamp in southern Arkansas, I rounded a bend and walked right onto a couple of moonshiners at their trade. They and I were equally surprised. I asked if they had seen any wild turkeys. They said no, gave me a hard look, and I hastily left. Nothing more happened. At another time, in Grand Bay, a swamp in southern Georgia, I passed each day for two weeks within a hundred yards of an active still, but was not aware of its presence until I read in the local paper that it had been raided by revenue agents. I have wandered onto many abandoned stills, particularly in the Great Dismal Swamp, which in the old days must have been a moonshiners' paradise.

The great Southern swamps and bottomlands occur on the Atlantic Coastal Plain north to southeastern Virginia and in the lower Mississippi Valley to extreme southern Illinois. Most of the swamps and bottomlands in the lower Mississippi Valley are in the Delta or Mississippi Alluvial Plain.

Swamps and bottomlands are characterized by relatively lush growths of hardwoods, often with shrubby and viny undergrowth, and by permanent or periodic flooding. Fifty percent or more of the forest is composed of oaks, gums, or cypress.

The terms swamp, river bottom, bottomland, hardwood bottom, and floodplain forest often are used synonymously. However, several of these terms refer to distinct physiographic units. The lowland forest bordering a Southern river is generally known to the forester or plant geographer as a river bottom or bottomland. Such an area usually is a complex of several forest communities, including swamps and floodplain forests. Swamps are permanently flooded except during droughts; thus they differ from floodplain forests, which are periodically flooded, usually in late winter or spring.

There are several types of swamps. Those occurring in bottomlands are known as river or alluvial swamps. They are found in the lowest part of the bottoms, either bordering the river or between the floodplain forest and adjacent uplands. Swamps found away from bottomlands are known as non-alluvial or inland swamps. Okefenokee is a good example.

The collection of photographs, experiences, and descriptions in this book is based on thirty years of travelling in swamps, river bottomlands, and canebrakes of the Southern United States. During some of this time I was conducting surveys as a biologist for the U.S. Department of the Interior. Additional time was devoted to studying the little known Swainson's warbler, conducting life history studies of other birds, and collecting plants.

Most of my time was spent in the Great Dismal Swamp in southeastern Virginia and northeastern North Carolina, the Ocmulgee River floodplain forest of south-central Georgia, the bottomlands of the lower Arkansas and White Rivers in Arkansas, Bayou Boeuf Swamp in central Louisiana, and the great Pocosins of eastern North Carolina. Briefer periods were spent in swamps and bottomlands of all of the Southern states.

Except for the sizes of the trees, some of the great swamps and bottomland forests probably appear today much as they did in earlier times. The vast tracts of lowland forest along the Santee River in South Carolina, the Altamaha in Georgia, and other Coastal Plain rivers of the Southeast cover hundreds of thousands of acres each. The Great Dismal Swamp wilderness covers nearly 500,000 acres and the Okefenokee more than 300,000.

Some of the swamps and bottomlands are denser now than before the virgin timber was removed. The disturbances caused by lumbering and burning usually results in a dense undergrowth. The few virgin tracts of timber that I have seen usually had a park-like appearance. The towering oaks and gums of the primeval forest are today found only in a few small tracts and as lone trees here and there.

Apparently the Singer Tract, lying along the Tensas River in northeastern Louisiana, was one of the last great virgin bottomland forests in the South. This tract contained 80,000 acres in 1937.[1] Ten years later the virgin timber was virtually gone.

The most magnificent virgin forest that I have seen was near Turkey Creek, Evangeline Parish, Louisiana. It was a second-bottom forest in which some of the oaks, hickories, and gums stood more than 125 feet tall. The lowest limbs were so high that I could identify many of the trees from the foliage only by using binoculars. This tract was completely leveled in 1957 by Hurricane Audrey, Louisiana's worst storm in the twentieth century.

The author in backwater or flooded area of Ocmulgee River bottoms, Ben Hill
County, Georgia, May 26, 1946. Flooding of Southern bottomlands usually occurs in
late winter or spring. (Photograph by Robert E. Gordon.)

Alluvial swamp along the Ocmulgee River, Ben Hill Country, southern Georgia, May 26, 1946. Principal trees are bald cypress and tupelo gum.

OKEFENOKEE COUNTRY

Okefenokee has all of the ingredients of the classic Southern swamp: the watery swamp forests, the live oak hammocks, alligators and large wading birds, and the legends. In my judgement it is the most picturesque swamp in North America, and certainly the best known.

Located in the southeastern corner of Georgia, it covers parts of Brantly, Camden, Clinch, Pierce, and Ware Counties, and dips a short distance into Florida. Four-fifths of the area is a national wildlife refuge that extends about thirty-eight miles from north to south and about twenty-five miles across at its widest part.

The great swamp is the source of the Suwannee and the St. Mary's Rivers. The Suwannee flows southwestward out of the swamp across northern Florida to the Gulf of Mexico; the St. Mary's proceeds eastward to the Atlantic, forming the Georgia-Florida boundary along most of its course.

The name Okefenokee, meaning "Land of Trembling Earth," was the apt and poetic designation of the Indians. Much of the swamp is a sort of floating peat bed that rocks back and forth where one attempts to walk on it, thus causing small trees and bushes to move up and down. The peat deposits range up to twenty-five feet in thickness. It is said that the big cypress trees are not in contact with the ground, but are rooted in the upper crust of the peat bed.

Dr. Francis Harper led the parade of eminent scientists, including several from Cornell University, that described and classified the fauna and flora of this primitive area. Dr. Harper, one of the last of America's trail-blazing naturalists, spent over 400 days in the swamp between 1912 and 1936, many with his colleague A. H. Wright. Their books and papers are natural history classics, and much of what we know about Okefenokee today is based on their work.

When Harper arrived at Okefenokee, logging operations were under way to remove the virgin cypress from the swamp. Logging continued sporadically until 1936, when most of this vast wilderness became a national wildlife refuge. Not all of the virgin cypress and gum were removed, and with the regrowth of timber in logged-over areas, Okefenokee for the most part looks today as it did when it was the stronghold of the Seminoles and Creeks.

Okefenokee is a mosaic of cypress bays, cypress heads, hammocks, shrub bogs, watery prairies, and lakes. Harper described these areas in detail in his book *The Mammals of Okefinokee,* published in 1927, as did Wright and Wright later in their monograph, "The Habitats and Composition of the Vegetation of Okefinokee Swamp," (1932).

A cypress bay may be described as an extensive wooded tract in which pond cypress is the dominant tree, with a lower layer of smaller trees, the bays: loblolly

13

Top: Francis Harper poling through Mud Lake, Okefenokee Swamp, May 23, 1912. (Photograph by David Lee.) *Bottom:* Francis Harper and guide David Lee in Hammock on Floyd's Island, Okefenokee Swamp, May 22, 1912. (Photograph courtesy of Dr. Harper.)

bay, red bay, and sweet bay. The undergrowth in this formation is usually composed of broad-leaved evergreen shrubs: hurrah-bush, ti-ti, and gallberry, and several less abundant species.

Harper describes cypress heads or "houses" as "wooded islets situated in the watery 'prairies' of the Okefinokee. . . . The prairie 'heads' are best known to the trappers and alligator-hunter, who find them convenient camping-places, and hence have come to refer to them as 'houses.' " [1] They differ little from cypress bays except in size.

Cypress bay, Cowhouse Island, Okefenokee Swamp, June 1970. Trees are pond cypress; understory mostly ti-ti, red bay, and sweet bay.

Left: Blooming loblolly bay or Gordonia in cypress bay, Suwannee Lake area, Okefenokee Swamp, June 1970. *Right:* Ti-ti probably is the most conspicuous shrub in Okefenokee when it is blooming in June.

The prairies are extensive flooded marshes with small ponds. Maidencane, a grass, is the common plant of the shallower parts. In deeper waters, the various pad plants or waterlilies, "neverwets," and "wampee" usually prevail.

Pine islands, some with hardwoods hammocks, have in earlier times been the main place of habitation for Indians and later for a few white families. Longleaf pine with an undergrowth of saw palmetto characterizes the dry pine barrens. The longleaf pine is replaced by slash pine in the more moist areas. Live oak hammocks occupy the higher parts of the islands.

16

Climbing heath, a shrub, is noteworthy for its unusual manner of growth. It works its way upward between the inner and outer layers of bark of the pond cypress, sometimes reaching a height of fifty feet. Billy's Lake, Okefenokee Swamp, June 1970.

Among Okefenokee's plant specialties are the fever tree or Georgia bark and the climbing heath. Okefenokee is about the center of the very restricted ranges of these two species.

The fever tree is a small tree attaining a height of up to about twenty-five feet. The most striking feature is its large rose-colored petal-like sepals, which measure 2 to 3 inches long and 1 to 1½ inches wide. In early times a decoction from the bark of the fever tree was used for treating victims of malaria.

From my field work in and around Okefenokee I would guess that the fever tree is rather rare. I have seen it more around the edges of the swamp than in it. It favors the borders of hardwood branch bottoms. In one such bottom or branch swamp near the Alapaha River some forty miles west of Okefenokee, I have seen it growing at the edge of a forest composed predominantly of tulip poplar, with an undergrowth of small trees and shrubs including the buckwheat tree, poison sumac, red maple, willow, and myrtle. In Okefenokee I have seen it growing in the drier sections of the pond cypress-slash pine association. During a visit to Okefenokee on June 20, 1970, I found the fever tree near the end of its flowering period, which is mostly during May.

The climbing heath is a vine that is noteworthy for its unique manner of growth. It makes its way up between the inner and outer layers of bark of the pond cypress, sometimes reaching a height of forty feet and sending out branches with leaves and flowers every few feet. The branches look as if they are growing right out of the trunk of the tree. It is an abundant plant in the Billy's Lake section of Okefenokee and perhaps other areas of the swamp.

Among Okefenokee's interesting fauna are the black bear, alligator, Florida sandhill crane, and the round-tailed muskrat. Despite the size and inaccessibility of much of the swamp, the local bear population is in danger of extirpation. Leonard Walker, biologist in the Okefenokee Wildlife Refuge in the middle 1960s, told me that there were only about 100 bears remaining. Many bears have been shot when raiding bee hives around the edge of the swamp.

The alligator lives along the watercourses and on the wet prairies and usually is near its special domain, the gator-hole. Francis Harper estimated the size of a gator-hole at fifteen to sixty feet in diameter and from three to five feet deep.[2] An occupied gator-hole is said to contain more fish than one in which no alligator

Gopher tortoise leaving den. In Okefenokee it occurs in the drier, sandy sections of hammocks.

Okefenokee alligator blending in with ripples and shadows.

is present. In periods of drought, gator-holes often become the main watering places of other denizens of the swamp.

In the winter of 1956–57, Eugene Cypert, Okefenokee Refuge biologist, estimated the Florida sandhill crane population of the swamp at 1,000 birds.[3] Before Okefenokee became a national wildlife refuge, Florida sandhill cranes were nearly extirpated from the swamp because of their high palatability rating with native hunters. One winter day in Okefenokee I saw a sandhill crane soaring with six turkey vultures at about 1,000 feet.

The round-tailed muskrat, or Florida water-rat, as it is sometimes called, reaches its northern limit in Okefenokee, where it is common. It is a small edition of the common muskrat, but is not related. Its body is eight inches in length, its tail five inches. This largely nocturnal mammal lives on wet prairies, where its small muskrat-like nests reach a height of only about six to ten inches above the water's surface.

Globular nest of round-tailed muskrat or Florida water rat, Okefenokee Swamp, June 1970. This small mammal is abundant here, near the northern limit of its range. (Nest near center of photograph, just above white waterlily.)

A black vulture, essentially a Southern bird, "sunning" in Okefenokee Swamp.
(Photograph by Luther Goldman.)

Water turkeys or anhingas in pond cypress, Okefenokee Swamp, Georgia, spring 1971. Water turkeys are primarily fish-eating birds that search for food beneath the surface of the water. (Photograph by Luther Goldman.)

In their list of the birds of Okefenokee, Wright and Harper in 1913 listed only one record of the limpkin and indicated that the ivory-billed woodpecker was an extremely rare bird, while the swallow-tailed kite and the Swainson's warbler were fairly common. There have been only two or three additional records of the limpkin from the swamp, where it would be north of the range of its main food, the apple snail. My friend Luther Goldman saw one there in the spring of 1971.

Records of the ivory-bill have been few since 1913. The last record was that of Hebard and Street, who saw a pair in Okefenokee on November 30, 1948.[4] Leonard Walker, Refuge biologist, told me that in the 1960s only about six swallow-tailed kites were seen each summer.

Writing on the status of the Swainson's warbler during their 1912 bird expedition in the swamp, Wright and Harper remarked that "To find that this famed and elusive warbler is not an uncommon inhabitant of the deep Okefinokee thickets, was one of the rarest pleasures of our sojourn in the swamp."[5]

In 1912, Wright and Harper found the crested flycatcher to be the most abundant songbird of the swamp. "This bird," they stated, "is found in every part of the Swamp and surrounding country. In fact only the red-bellied

Billy's Lake, Okefenokee Swamp, Georgia, June 1970.

The oak toad of Okefenokee hammocks is known by its baby chick peep-like whistle, and the white line down the middle of its back. This elfin toad is only about an inch long.

woodpecker can compare with it in numbers and in widespread distribution."[6] In a one-mile census along Billy's Lake on June 20, 1970, I likewise found the crested flycatcher to be the most abundant songbird. The prothonotary and parula warblers, yellow-throat, red-winged blackbird, and Carolina wren were next in abundance in that order. The little blue heron was the most common large wader.

The most recent important ornithological discovery in Okefenokee concerns a nesting colony of wood ibis. This species was thought to nest only in southern Florida. On July 5, 1967, William C. Cone and Jewett V. Hall, employees of the Refuge, found a rookery on Craven's Hammock, an island which is rarely visited and virtually inaccessible except during periods of low water. There were a dozen nests in the rookery, each with large young.[7] A record of the scarlet ibis in the spring of 1971 was the first for the swamp and the state of Georgia.

In my various sojourns in Okefenokee, I suppose that nothing has attracted my attention more than the frogs and toads. There are at least twenty-one species in the swamp. In June 1970, I probably heard most of them, from the largest, the raucous-sounding Southern bullfrog or pig frog to the tiniest, the little grass frog. Each night at about eleven o'clock, I was serenaded by the midget of American toads, the oak toad. It calls from the piney islands and live oak hammocks of Okefenokee, as well as the surrounding southeastern coastal pine flatwoods. This elfin toad is 1¼ inches in length, and is blackish with a whitish line down the middle of its back. Its voice is a high-pitched whistle that sounds like a baby chick peeping.

24

This striking photograph of white ibis and a few little blue herons was made by
Luther Goldman in Okefenokee Swamp, Georgia, spring 1971.

The swamp's list of toads and frogs includes the following:

Eastern Spadefoot Toad	Southern Chorus Frog
Southern Toad	Ornate Chorus Frog
Oak Toad	Eastern Narrow-mouthed Toad
Southern Cricket Frog	Bullfrog
Southern Spring Peeper	River Frog
Green Treefrog	Pig Frog
Pine Woods Treefrog	Carpenter Frog
Squirrel Treefrog	Bronze Frog
Eastern Gray Treefrog	Southern Leopard Frog
Barking Treefrog	Florida Gopher Frog
Little Grass Frog	

For a most fascinating account of the frogs of this great swamp, one should read *Life-Histories of the Frogs of Okefinokee Swamp, Georgia* by Wright (1932).

Just as interesting as the plant and animal life are the people who have lived, worked, hunted, and trapped in the swamp. The story of these people would fill a number of book-length volumes. The Lees of Billy's Island, the Chessers of Chesser's Island, the Mizells, Lem Griffis, and others knew the wildlife of the swamp from the point of view of the hunter and trapper.

John M. Hopkins, superintendent of logging operations from 1900 until the area became a national wildlife refuge, worked in the swamp for forty-five years. He also was Okefenokee's first Refuge manager. In his book *Forty-five Years with the Okefenokee Swamp* he makes this interesting statement in expressing his appreciation of the place: "In the years I have known it and loved it, I have found the great swamp far from being a place of mystery, danger and menace, but rather a haven of peace and a refuge from the greater hazards of the outside world."

THE GREAT DISMAL SWAMP

IN TRAVELLING southward along the Middle Atlantic Coast one may wonder where the North ends and the South begins. Nowhere is this distinction more marked than in the region of the Great Dismal Swamp; and in no way is it better expressed than in the flora and fauna of the area.

Because of the moderating influence of the Gulf Stream, which follows the coast northward to about southeastern Virginia before bending out to sea, many of the physical features of the Dismal Swamp are like those of the swamplands of the deep South. Many kinds of plants and animals primarily associated with the lower South are "pushed" northward along the coast by the warm littoral climate, reaching their northern limit in this region.

This 500,000-acre wilderness stretches from a few miles south of Norfolk, Virginia, to the vicinity of Elizabeth City, North Carolina. Lake Drummond, the principal focal point of the swamp, is located on the Virginia side. The lake is one of the highest points in the swamp at about twenty-two feet above sea level—a sort of inverted saucer. Its dimensions are about 2½ by 3 miles, its greatest depth is about fifteen feet, and its amber-colored water is like that of other Southern cypress-gum swamps.

Numerous canals, some of which were dug by slave labor, radiate in all directions from the lake to the upland edge of the swamp. One of these ditches was dug under the direction of George Washington in 1768. North Jericho Ditch, located in the northwest section near Suffolk, has the finest forest in the swamp, is the most varied botanically, and has the best assortment of birds.

Despite its nearness to a great urban center, and constant lumbering and drainage for 200 years, the swamp is still a vast wilderness, even though most of it is second-growth forest.

Because of the many attempts to drain the swamp it is drier than formerly, which has led to the invasion in some sections of plants requiring a drier substrate. Such upland plants as white oak are occasionally found one or two miles out in the swamp. But in spite of man's interference the Dismal is still mostly a wetland area. Compared to Okefenokee, the Dismal is more of a solid forest, while Okefenokee is a complex of forests, lakes, and grassy prairies.

Most of the soils are saturated and very acid. The deep water swamp, characterized by cypresses and gums, has a humus soil, while the Atlantic white cedar forest and the evergreen shrub-bog community, with less surface water, are underlain by peaty soil deposited in some places to a depth of ten feet.

The hydric or deep water swamp forest is the wettest in the swamp. In some places plants may be growing in two feet of water. These same plants, however,

Left: Supplejack or rattanvine near northern limit of range, in the Great Dismal Swamp, Virginia.

Below: Yellow jessamine near its northern limit, in the Great Dismal Swamp, southeastern Virginia. The fragrance from the yellow flowers of this evergreen twining vine pervades the Dismal in middle and late April.

Paw paw fruit and leaves. Paw paw belongs to the custard-apple family, and is edible. I found this small, widely distributed tree in the Dismal Swamp, Virginia; Bayou Boeuf Swamp, Louisiana; in a river front hardwoods forest along the Arkansas River, Arkansas; and in many other Southern lowland forests.

may grow on drier sites. The type is best characterized by bald cypress, tupelo gum, swamp black gum, and red maple. Bitter pecan (rather uncommon in the swamp), water ash, and swamp poplar sometimes occur with the gums and cypress. Shrub strata plants such as Virginia willow and buttonbush occur on less wet sites. The largest cypress that I have seen in the swamp has a diameter at breast height of five feet three inches.

The most diverse collection of plants occurs in the semi-hydric or mixed swamp hardwoods forest. This community apparently evolved from a wetter, pure stand of swamp black gum forest as the swamp became drier. Most of the year the mixed swamp hardwoods forest is without standing water, but is damp.

This forest type is mainly composed of swamp black gum, sweet gum, red maple, water oak, swamp magnolia, red bay, American holly, and paw paw. Sweet pepperbush often is the major understory shrub. Rattanvine or supplejack, virtually at its northern limit, and greenbrier frequently occur with sweet pepperbush, forming dense tangles. In the more open parts of the forest the netted-chain fern is a conspicuous ground cover plant; and two Southern lianas, the fragrant yellow jessamine and the showy climbing hydrangea, grace the trees.

The mesic or hammock forest, driest in the swamp, is a sort of wooded island. Its common trees are white oak, swamp chestnut oak, cherrybark oak, beech, tulip poplar, hornbeam, and American holly.

29

It was in a small tract of hammock forest along the north end of Jericho Ditch that I found two of the rarest plants in southeastern Virginia, the wild or silky camellia and the dwarf trillium. The wild camellia, a small tree in the same family as Franklinia and Gordonia (Theaceae), reaches a height of about fifteen feet, and has large showy blossoms with yellow stamens, that appear in late May. The dwarf trillium grows to a height of only one to three inches above the leaf litter, and blooms in late March and early April. Its flowers turn from white to purple.

A short distance down Jericho Ditch from the trilliums and wild camellias, where the woods are a little damper, I found a hybrid azalea, a cross between the swamp azalea (*Rhododendron viscosum*) and the pink azalea (*Rhododendron nudiflorum*). Some of these plants reach a height of twelve feet. The white flowers appear in late April and early May.

The extensive Atlantic white cedar stand of the Dismal has the appearance of a coniferous woodland of the far North. It is the largest stand in existence. Because of the high value of this timber and the extensive remaining stand, it was still being logged in the early 1970s. Undisturbed stands are so dense that the understory shrubs are few; but in less dense areas highbush blueberry and shining inkberry are typical.

Dwarf trillium in the Dismal Swamp, Nansemond County, Virginia, April 1970. The flower appears to change from white to pink to purple. It is one of the earliest herbaceous plants to bloom in the Dismal.

Flowers of the wild camellia or Stewartia in the Dismal Swamp,
Nansemond County, Virginia, May 1969. This small tree is rare
in southeastern Virginia, where it reaches its northern limit.

Where Atlantic white cedar stands have been logged off or burned, they usually
have been followed by an evergreen shrub-bog community. This complex of
predominantly broad-leaved evergreen shrubs, bay trees, and pond pine reaches
its northern limit in the Great Dismal. Such relatively open areas in the swamp
are known as "lights." They have the same green appearance the year round.
Principal broad-leaved evergreens of the community are ti-ti, fetterbush, downy
leucothoe, red bay, gallberry, and laurel smilax. The first four are definitely
Southern.

Left; Trunk of bald cypress in the Great Dismal Swamp, North Carolina, April 1971. Diameter at breast height is 5′3″.

Below: Pond in Atlantic white cedar peat bog, Dismal Swamp, July 1971. Eggs of four species of frogs were in this pond when photograph was made.

Atlantic white cedar along Virginia-North Carolina boundary in the Great Dismal Swamp, approximately twenty miles southwest of Norfolk, Virginia, January 19, 1961.

Each of the swamp's major plant communities has its animal specialties, but many species of birds, mammals, reptiles, and amphibians cross lines, occurring in the same habitats. Local conditions usually determine the distribution of these forms. I found the pine woods treefrog and little grass frog commonest in the evergreen shrub-bog community; the squirrel treefrog mainly in the mixed swamp hardwoods forest; the spotted turtle in the tupelo gum swamps; the yellow-bellied turtle in the larger canals, and so forth.

The presence of black bears, bobcats, white-tailed deer, and river otters give the swamp a wilderness aura. The sight of a bear in the depths of this forest is a thrilling experience. The Dismal Swamp is about the last place that native bears occur in the Middle Atlantic Coastal region. Apparently there were fewer than 100 bears there in 1970.

In late June 1971, I was walking along the logging road that borders Jericho Ditch when a yearling bear emerged from the swamp forest undergrowth about 200 feet ahead of me. Apparently it did not see me and wandered leisurely down the road for about another 200 feet before heading back into the forest. It must have been on the way to its honey tree, because at the point where it entered the forest, hundreds of honeybees, other kinds of bees, and various flies were buzzing around a mixed clump of sweet pepperbush shrubs and swamp magnolia saplings. A number of honeybees were stuck to the foliage which had a glazed appearance. I touched one of the leaves with the tip of my tongue and identified the substance as wild honey. Later, I saw pieces of honeycomb lying on the ground nearby.

Logging operation, Great Dismal Swamp, June 1971.
The large log in center of photograph is a cypress.

Bear honey tree, Dismal Swamp, June 1971. Arrow points
to hollow in bro[ken] [lim]b where honey is located.

Looking upward I saw bees swarming about the hollow of a large broken-off
branch, which was about fifty feet above the ground (see photograph above).
Claw marks of the bear were evident on the trunk of the huge tulip poplar in
which the bee gum was located. To obtain the honey the bear would reach into
the hollow, and as it withdrew its paw, honey, pieces of comb, and bees that
had been crushed fell below onto the foliage of the undergrowth.

Along the trail to the honey tree the yearling bear left its "calling card" and
from this I was able to find out more about its food habits. I took a sample of
the scat (mammalogist's term for animal feces) to the laboratory of the Patuxent
Wildlife Research Center, where I examined the material under a microscope.
In addition to a few blueberries, the bear had been feeding mostly on insect larvae

Swamp magnolia, Great Dismal Swamp, in May.

or grubs which it had taken from a rotten log. The source of the grubs was quite evident as most of the material in the scat consisted of pieces of rotten wood which were ingested along with the food items.

The river otter is still a common mammal in the swamp. It is mainly associated with the ditches or canals. In April 1968, I observed a pair with two pups, about two-thirds grown, at their den in a bank of the Jericho Ditch. I could always tell when the otters were in the den as I could see "steam" pouring out of both entrance holes, which were about two feet apart at water level in the ditch bank.

There was a lone otter operating in the same section of the ditch in June 1969. It did most of its feeding in the evening, beginning about 5:00 P.M. Since there is a trail beside Jericho Ditch it was possible to follow the otter as it swam underwater searching for food. The Jericho Ditch trail was about four feet above the water level. Once I followed the otter for three-fourths of a mile. As it swam along under the water a string of bubbles trailed behind on the surface. Each time it caught something (usually crayfish) it would surface and, with head and neck above the water level and nose pointed straight up, it would chew its food for a few seconds, then dive and resume foraging.

The Dismal Swamp is the type locality of one mammal species and three subspecies. The Dismal Swamp short-tailed shrew (*Blarina telmalestes*), discovered by A. K. Fisher in 1895, is an endemic species. Bachman's shrew (*Sorex longirostris fisheri*), the Southern bog lemming (*Synaptomys cooperi helaletes*), and a muskrat (*Ondatra zibethica macrodon*) are the subspecies. The golden mouse, also found in the Dismal Swamp, was discovered a few miles away at Norfolk. The marsh rabbit is one of several Southern mammals that reaches the northern limit in the Dismal. I have seen only one in fifteen years of exploring in this great swamp.

There is a rich bird fauna in the swamp, but not many so-called specialties. There are about eighty species of breeding birds, including fifteen species of warblers.

In the mixed swamp forest of North Jericho Ditch, the five most abundant breeding birds in the late 1960s were the prothonotary warbler, red-eyed vireo, hooded warbler, Wayne's black-throated green warbler, and ovenbird. On the basis of my studies in Southern swamps and bottomlands, I would say that the Wayne's warbler and the swamp's other Southern bird specialty, the Swainson's warbler, probably are more abundant in the Great Dismal than elsewhere in their breeding ranges.

The Wayne's warbler, geographic race of the black-throated green warbler, is found during the breeding season along the South Atlantic Coast from the Dismal Swamp near Norfolk, Virginia, to Charleston, South Carolina. (Drawing by John W. Taylor.)

The Wayne race of the black-throated green warbler is the second resident warbler to arrive in the swamp in the spring. It follows the yellow-throated by only a few days, some individuals arriving by the last week in March. Since it is one of the earliest resident warblers to appear, it is one of the first to nest. Apparently it starts nest building before the better known Northern race of the black-throated green has left its wintering ground in tropical America! There are records of nests and eggs in the Dismal Swamp as early as the first week in April.

On April 12, 1969, I observed a female gathering nesting material. She was taking the wool-like coating from the stems of young cinnamon ferns. The coating is used in the lining of all nests of this warbler that I have found in the Dismal.

While the Wayne's warbler is a high ranging species, the little known Swainson's warbler lives mostly on the ground and in the undergrowth. Most of the nests that I have found in the Dismal were a few feet from the ground in tangles of sweet pepperbush and smilax. All nests were lined with the pedicels of red maple fruits, like those I found in the Ocmulgee River canebrakes in Georgia.

It seems strange that the Kentucky warbler is absent from the swamp as a breeding bird; that the parula warbler is restricted mostly to cypress during the breeding season; and that the prairie warbler occurs as a nesting associate of the Swainson's, prothonotary, and hooded warblers in closed forest habitat of the mixed swamp hardwoods community.

The Atlantic white cedar forest is the most sterile habitat for breeding birds in the swamp. The blue jay and wood thrush seem to be the only breeding birds in dense stands. In winter the winter wren, a visitor from the North, is one of the few birds that occur in the dark understory of dense stands of cedar. One winter I estimated that 10,000 pine siskins were feeding on seeds of the Atlantic white cedar along Corapeake Ditch.

The most extraordinary sight in the Dismal is the winter blackbird roost in the evergreen shrub-bog area along the Carolina-Virginia boundary. It is the largest roost in the country with an estimated 30 million birds. Red-winged blackbirds, common grackles, brown-headed cowbirds, rusty blackbirds, and starlings make up this aggregation. An estimated one million robins also winter in this same section of the swamp.

The Dismal Swamp is considered to be good snake country, but one can go for several days in spring and summer without seeing any. In late June 1971, I spent four days in the Dismal, and the only snake that I saw was dangling from the talons of a red-shouldered hawk.

Three poisonous snakes—the copperhead, cottonmouth, and canebrake rattler—occur in the swamp. The last two are at about their northern limit. The mud snake also reaches its northern limit in this region. My colleague Francis M. Uhler has seen the scarlet snake a few miles from the Dismal.

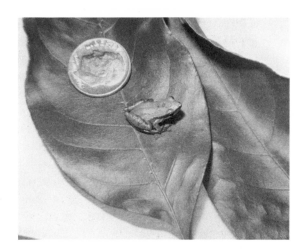

In the Great Dismal Swamp of southeastern Virginia, I found the little grass frog in the evergreen shrub-bog community and elsewhere in temporary rainwater ponds. (Photograph by Jerry Longcore.)

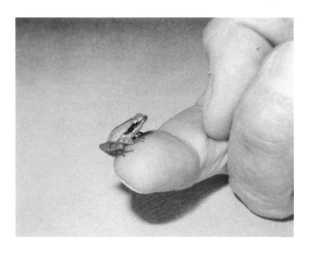

Adult little grass frog, tiniest North American frog, reaches its northern limit in the Great Dismal Swamp region of Virginia. (Photograph by Jerry Longcore.)

Brimley's chorus frog is found along the Outer Coastal Plain from the Dismal Swamp, in southeastern Virginia, to about Savannah, Georgia. This specimen was found in the Dismal Swamp in September 1971. (Photograph by Jerry Longcore.)

The little grass frog, tiniest frog in North America, the squirrel treefrog, and Brimley's chorus frog are three Southern species, at or near their northern limit. All three are locally common. On June 29, 1971, I heard nine squirrel treefrogs calling in a one-acre area in the woods along North Jericho Ditch. I have found the little grass frog, Southern cricket frog, and Southern toad calling from a rainwater pond at the head of Railroad Ditch in late March and early April.

On April 24, 1971, I spent a pleasant morning with Robert T. Mitchell, an entomologist, collecting butterflies along North Jericho Ditch. We found four species of swallowtails, the palamedes, pipevine, tiger, and zebra; also Juvenal's dusky wing, silverspot, gemmed satyr, orange tip, common blue, and Creole pearly eye. We were especially interested in the zebra, which is common here because of the paw paw upon which its larvae feed. At this time the adult zebras were feeding to a great extent on the nectar of flowers of downy leucothoe. The palamedes is one of the most abundant and widely distributed butterflies in the swamp. The spicebush swallowtail was common along Jericho Ditch in June.

Many kinds of butterflies, frogs, birds, and other fauna and plants are protected by this northernmost of the great Southern swamps of the Atlantic Coastal Plain. But like any large undeveloped tract of land located next to an urban area, the Great Dismal Swamp is vulnerable to exploitation. Unfortunately its half million acres are not an established sanctuary like the Southern Everglades and Okefenokee. It is hoped that a part of this scientifically important and historic area can be preserved.

THE LOWER ALTAMAHA

THE GREAT river swamp of the lower Altamaha, near the Georgia coast, is best known to the botanist as the locality where John Bartram and his son William discovered the Franklin tree, better known by its generic name Franklinia. The type locality is given by the Bartrams as near Fort Barrington, the site of which lies on the north shore of the Altamaha River, some fifteen miles upstream from Darien.

Franklinia was discovered in 1765, and twenty-five years later was extinct in nature. It exists as a cultivated plant today because of seeds taken by William Bartram to his botanical garden in Philadelphia. The plant must have been rare at the time of its discovery since it was never known to exist elsewhere.

Its extinction at the only known natural site is thought to have resulted from plants removed by collectors and shipped to nurserymen in England, where there was a great demand during the 1700s for rare and unusual plants from America.

The Bartrams named this attractive plant in honor of Benjamin Franklin, and for the river near where it was discovered, giving it the scientific name *Franklinia alatamaha*. The original spelling for the river included an extra *a*. Franklinia is in the same family (Theaceae) as Gordonia or loblolly bay and Stewartia, the wild camellia.

It is a small tree, reaching a height of fifteen to twenty feet. Its flowers are white with yellow stamens, and are about three inches in diameter. Trees at my home in Maryland bloom in August and September.

The Bartrams found the Franklinia growing in a three-acre sandhill bog or branch swamp not far from the site of the old fort. Dr. Francis Harper, biologist and author of several books on the Bartrams, reported the following interesting mixture of plants at a wet site near Fort Barrington, some of which would probably have been associated with Franklinia: pond pine, red bay, Georgia bark or fever tree, hurrah-bush, and laurel smilax.[1] Those of us who have searched for Franklinia in its natural state have found only its close relative, the loblolly bay, that grows abundantly on damp sites near Fort Barrington.

A few miles down the river from Fort Barrington, the swamp begins to thin out and merge with the marshlands. It is in this transition or ecotone area that I occasionally see one of the most spectacular and graceful of our avian aerialists, the swallow-tailed kite, a rare bird in this region. Ivan Tomkins, the Georgia ornithologist with whom I surveyed the Altamaha swamp, published the following note about an unusual encounter:

On the morning of 10 May 1964, Mr. Bailey Darley, who operates the fishing camp at the South Altamaha River, just off U.S. Highway 17, in McIntosh County, Georgia, saw a bird, unfamiliar to him, alight on the road nearby. The bird was fluttering one wing. He found that the bird had a small snake wrapped around the wing. With the aid of a stick he was able to get bird and snake into a wire poultry cage. He then reached in with a stick and got under the snake, which released its hold. Through Mr. George Geiger of the Georgia Fish and Game Department, the word was passed on to Mr. Neil Hotchkiss, of the Patuxent Research Center, and he and Mr. Walter A. Harmer went over that day, and identified the bird as a Swallow-tailed kite (*Elanoides forficatus*) and the snake as the keeled green snake (*Opeodrys aestivus*).

The following day, Mr. Brooke Meanley, also of the Patuxent Research Center, and I saw the bird and the reptile, and photographed the bird. It was an adult in excellent plumage other than a minor amount of feather damage from the incident. Mr. Darley

Site of old Fort Barrington, Altamaha River, Georgia,
March 1971. This is the type locality of Franklinia.

Blossom of *Franklinia alatamaha*. The Franklin tree was discovered by John and William Bartram at Fort Barrington on the Altamaha River, Georgia, in 1765. Twenty-five years later it was extinct in the wild.

released the bird and the snake that same day. When released the kite flew toward the top of a nearby cypress, but was chased by two Mockingbirds and flew on off with no obvious damage.

Recent records of this kite in coastal Georgia are very few. Mr. Harmer, who is a biologist with the Georgia State Board of Health, has spent many years in the Altamaha River Delta and has never seen the species there before.

It is supposed that when the kite captured the snake in its claws, it wound around the wing enough to hamper flight. The bird was quite docile in captivity and accepted and ate a few small minnows offered to it on a straw or wire.[2]

Apparently the swallow-tailed kite was formerly much more numerous and widely distributed in the United States. Its breeding range used to extend up the Mississippi Valley into Minnesota and Wisconsin. Today, this kite is fairly common only in the Everglades region of south Florida. It has become extinct in most of its former U.S. breeding range.

At Augusta, Georgia, just below the Fall Line, where the swamps and floodplain of the Savannah River begin, E. E. Murphey in 1937 reported an estimated 100 kites soaring in the air one September day in 1893. A swallow-tailed kite reported at Augusta on July 30, 1967, was the first record of that species from that area since May 1943.

There is only a handful of recent records from the lower Altamaha and Savannah. In the 1960s I saw several in Monkey John Swamp near the city of

Swallow-tailed kite, rare bird of the Altamaha
River delta. (Illustration by John W. Taylor.)

Savannah. Two nests, each with two young, were located in the Santee River
delta in the South Carolina Low Country, May 17, 1961, and June 12, 1961.[3]
Recent sightings usually have been of swallowtails coursing with Mississippi kites.

Dr. Murphey made an interesting observation regarding an apparent change
in the color of the plumage of these birds shortly after their arrival on the breeding
grounds just below Augusta, Georgia:

> On arrival the breast and belly feathers were immaculately white, but in a short time
> thereafter they were stained on belly, flanks, and crissum (under-tail coverts), an
> ineradicable tobacco-juice brown due to the spurting of the body fluids of the grasshoppers
> crushed in their talons, which stain remained until the next moult.[4]

44

CANEBRAKES OF THE OCMULGEE

T HE Ocmulgee River drains the heartland of Georgia. Flowing south through Macon, it gradually bends eastward and joins the Oconee near Lumber City. The confluence of these two rivers forms the beginning of the Altamaha which reaches the coast a few miles east of Darien.

This great river system is notable for its broad bottomland forests which border the river along most of its course. In some sections of the Altamaha, the bottomland forest is five miles wide.

Aerial view of riverbottom wilderness along the Ocmulgee River, Wilcox County, south-central Georgia, autumn 1970. A few miles downstream, the Ocmulgee meets the Oconee to form the Altamaha. (Photographs by Milton N. Hopkins, Jr.)

For the first five miles or so south of Macon, the Ocmulgee bottomland is dominated by a floodplain forest of mixed hardwoods and tupelo gum swamps. This area along the river is one of the few stations in the Georgia Coastal Plain where I have seen silver maple. Hackberry, elm, ash, sweet gum, box elder, and swamp chestnut oak are important trees of the Ocmulgee floodplain forest.

Canebrakes are the most striking feature of this upper Coastal Plain section of the river. This is the giant species (*Arundinaria gigantea*). It does not grow in the tupelo gum swamps, but in the drier floodplain forest. Giant cane occurs in scattered patches from less than a mile south of Macon to about twenty-five miles down river. In the early 1940s, I saw a nearly pure stand covering a one-square-mile area. By the 1970s, the Ocmulgee River canebrakes were still the most extensive of any that I could locate in the South.

The canebrake is one of the most distinctive understory plant formations of the bottomland forest. This plant formation is unique because its life-form is unlike that of the usual shrubby or brushy undergrowth.

Cane is the only native American bamboo. Two species are known: giant cane and switch cane. Giant cane of the Ocmulgee River floodplain forest reaches a maximum height of about thirty feet, with an average of fifteen to twenty feet. In 1778 the botanist William Bartram recorded cane poles in the Tombigbee River bottoms above Mobile, Alabama, thirty to forty feet in height and three to four inches in diameter.[1] I have never found poles more than 1¼ inches in diameter. The maximum height of switch cane is about ten feet, with a diameter up to about ½ inch.

Travelling in a canebrake is like walking through a forest of fishing poles, and since everything looks the same in all directions, a tract of cane is a likely place in which to get lost. Canebrakes were often mentioned in the literature of the pioneer naturalists and hunters because they were one of the best places to find wildlife. Such a stand of native bamboo forms a dense, tall, uniform understory in which an animal can move about with relative ease and remain well hidden at the same time. Audubon's most famous painting, the Wild Turkey Cock, was based on his observation of one of these large birds in a Louisiana canebrake.

The tender young shoots of cane are utilized for food by the swamp rabbit, white-tailed deer, and some rodents. And since canebrakes often form the understory of sugarberry, sweet pecan, and oak forests, they provide excellent foraging areas for such game animals as turkeys, deer, and squirrels. The variety of wildlife that occurs in canebrakes is another indication of the importance of this habitat.

Opposite: Canebrake (cane species is *Arundinaria gigantea*) in Ocmulgee River floodplain forest, approximately five miles south of Macon, Georgia, April 1967. Greatest diameter of cane poles is about 1¼ inches; average height about eighteen feet. Tallest pole in photograph is thirty feet. Typical habitat of the Swainson's warbler.

Left: Nest and eggs of Swainson's warbler in canebrake near Macon, Georgia. Swainson's is one of the few warblers that has pure white eggs. *Right:* Black vulture "nest" and eggs on ground in canebrake along the Arkansas River, March 1, 1952.

The Swainson's warbler, one of the least known of Southern songbirds, is more closely associated with canebrakes than is any other bird. I do not recall ever being in an extensive stand of giant cane during the nesting season without seeing this bird. Although relatively common in some areas, it is a difficult bird to find because of the character of its habitat, its neutral color, and its habit of spending much time close to or on the ground in the shade. Thus it would rarely be observed were it not for its song, one of the most outstanding of warbler songs.

In the canebrake habitat, Swainson's warblers usually occur in small groups or "colonies" such as are characteristic of Kirtland's warblers on their highly restricted jack pine breeding ground in Michigan. Hence, it is often possible to hear three or four singing Swainson's males at one time.

In the Ocmulgee River canebrakes, these warblers arrive soon after April 1. Eighty-seven of ninety-one territorial males that I observed in eight nesting seasons three to five miles south of Macon had their territories (each averaging a little less than one acre) in patches of cane growing beneath the forest canopy. There

Young great horned owl along the edge of the Ocmulgee River bottoms, Wilcox County, Georgia. In the deep South, horned owls are associated more with piney woods, but sometimes hunt in swamps.

were about 20,000 cane poles per acre in one of my study areas.[2] The ground was virtually bare due to the shading effect of the upper and lower canopies, the cane understory, and periodic flooding.

Swainson's warbler nests in canebrakes are rarely located in the denser parts of the stands, but are nearer the edges where the stands are thinner and the cane poles are smaller. For some three or four days prior to nest building, one of the activities of the mated pair is the examining of nest sites.

Nest building begins in early May. The Swainson's warbler builds a large, bulky nest, larger than that of most other North American warblers that nest above the ground. The nest, usually placed at a height of two to five feet, is built by the female from materials gathered nearby. She works generally from 7:00 to 11:00 A.M., completing the nest in two to four days. Three or four white eggs comprise the clutch.

In a canebrake the foundation of a nest is often a bunch of dead leaves that has lodged in a cane stalk. The cup of every nest that I have found in the Ocmulgee canebrakes and in the Dismal Swamp was lined with the pedicels or stalks of red maple fruits.

The song of the Swainson's warbler is loud and ringing and of marked musical quality. As Edward Von Siebold Dingle states,

> The bird student who hears the song of Swainson's Warbler as he sings in his wooded retreat is fortunate, for it is one of the outstanding warbler songs and, once heard, leaves a lasting impression upon the listener. At a distance it bears much resemblance to the songs of the hooded warbler and Louisiana waterthrush. Close up, however, the appealing quality, lacking in the other two, impresses the listener strongly.[3]

Swainson's warblers sing from the ground and from vines, branches, and logs, usually below thirty feet. The song consists of three or four sharp, high introductory notes, all well separated, followed by a phrase of four or five syllables uttered rapidly and slurred.[4] Some birds seem to say, "See, see, see, what's in here."

I spent a June day in a male Swainson's warbler territory in the Great Dismal Swamp to determine the number of songs he sang from the time he started singing

Canebrake rattler, Coastal Plain form of the timber rattler, inhabits Southern canebrakes and other bottomland habitats. (Photograph by Ivan R. Tomkins.)

Wood rat nest in canebrake near Macon, Georgia. Two of cane poles in right foreground are about one inch in diameter.

at dawn, until he stopped for the day in late afternoon. The bird started singing at 4:27 A.M., and sang 1,168 songs before stopping at about 5:00 P.M. Most of the singing took place prior to 9:00 A.M. When singing his usual course or series of songs, he sang at the rate of about six songs per minute. Only the cardinal, wood thrush, and wood pewee began singing earlier in the morning and sang later in the evening than did the warbler.

Swainson's warblers do most of their hunting for food on the ground. Unlike other ground-feeding warblers which obtain much of their food from the surface of the leaf litter, the Swainson's searches for food mainly by shoving up leaves and looking on the underside or searching the ground beneath.

It was once thought that Swainson's warblers were found only in canebrakes. Fortunately this is not true, for the great canebrakes of the Southern bottomlands are rapidly disappearing due to cattle grazing off the young growth and the clearing of the cane for farm land.

THE WHITE RIVER WILDERNESS

THE bottomland forest along the lower White River in southeastern Arkansas is one of the last great wilderness areas in the lower Mississippi Valley. It is a remnant of a once unbroken stretch of bottomland hardwoods that extended from Cairo, Illinois, to below New Orleans. The White River wilderness remains largely as a vast tract of 117,000 acres, maintained by the U.S. Department of the Interior as a national wildlife refuge. There also are considerable areas of forestland extending north and south of the boundaries of the Refuge that are still good buffer zones, and contiguous to the White River bottomland forest is the remnant wilderness bottomland of the lower Arkansas River.

The White and the Arkansas enter the Mississippi only about ten miles apart. The tract of land separating these two rivers where they run into the Mississippi is known as Big Island. Big Island fronts on the Mississippi and is separated from the mainland on its other side by a channel known as the Cut-Off. Big Island is one of the most isolated and wildest sections of land along the Mississippi River, and during Prohibition it figured prominently in the whiskey trade.

I lived near the White River wilderness for five years, 1950–55, and later made numerous trips there until 1970. In the early 1950s I explored some of this great wilderness with Peter J. Van Huizen, manager of the White River National Wildlife Refuge. It was necessary to use a boat to reach the more isolated sections of the area. One of Pete Van Huizen's assistants was Lloyd MacAdams, a native who grew up as a hunter and trapper in the White River bottoms. A few years before the Refuge was established in 1935, he located the nest of an ivory-billed woodpecker in the bottomland. I had many pleasant experiences travelling through the White River wilderness with these two naturalists looking for bear sign, exploring canebrakes, and always on the lookout for the ivory-bill, which apparently was extirpated in this area by the late 1930s.

The lower White River wilderness is nearly fifty miles long, and in places five miles wide. It is interlaced with bayous, chutes, and channels that connect with the river. There also are over 200 lakes ranging up to 100 acres in size. Many of these are cypress-studded old riverbed or oxbow lakes.

The most remote parts of the White River wilderness are Scrub Grass Bayou and East Moon Lake, located in the southern part of the Refuge. This is where most of the bears occur. Also, it is in the canebrakes of this area that the Swainson's warbler, rarest songbird of this wilderness, is most numerous.

The bottomland is characterized by excessive flooding, usually in late winter and spring. Parts of this bottomland often are under twenty-five feet of water, and 75 percent of it may sometimes be inundated. The flooding of the bottomlands

Bald cypress, Oxbow Lake, White River bottoms, southeastern Arkansas.
Circa 1956. (Photograph by Peter J. Van Huizen.)

A former giant of the lower White River bottoms. A black vulture had a nest in the hollow of this sycamore. (Photo by P. J. Van Huizen.)

may have a devastating effect on nesting turkeys if it occurs in the spring; however, in late winter this inundation of the bottomland makes the acorn mast available to the hordes of wintering mallards.

The most striking feature of this great wilderness is its magnificent forest. Although some of the area has been logged, many giants of the forest still remain. I have measured a number of trees with diameters at breast height of six feet and several that were more than 125 feet in height.

In this bottomland forest, elevation and soils are the principal physiographic features that determine characteristic plant communities or forest types. Most of this great wilderness forest grows on heavy, poorly-drained alluvial clay soils.

In low, poorly-drained flats the overcup oak-bitter pecan type is predominant. This forest type is often flooded for protracted periods. The sweet gum-water oaks type is found in much of the better-drained parts of the first bottoms. Characteristic species of this association are sweet gum, water oak, Nuttall oak, willow oak, American elm, sugarberry, green ash, red maple, and cedar elm. Plants

54

Bear den in overcup oak. Deep overflow, Scrubgrass-East Moon Lake area, White River bottoms, Desha County, Arkansas. Circa 1956. My friend Peter Van Huizen, who took this picture, climbed up to the den, but found the bear not at home. The den was warm, however; Van said that he was glad the bear wasn't there.

commonly found in the shrub strata are swamp privet, deciduous holly, and haw. Common vines include greenbrier, grape, supplejack, Virginia creeper, peppervine, poison ivy, and buckwheat vine.

Sweet gum, sweet pecan, and Southern red oak usually are predominant on well-drained first bottom or cane ridges. Nuttall oak also occurs on these higher sites. Bald cypress, tupelo gum, planer-tree, and water locust are characteristic trees of bayous and sloughs that flow into the White River.

Nuttall oak, sweet and bitter pecan, and buckwheat vine are particularly identified with this wilderness area. Nuttall oak, named for the New England naturalist Thomas Nuttall, who visited the White River wilderness in the 1820s, is restricted to the alluvial bottomlands of the lower Mississippi Valley.

The logs of the Nuttall oak are known to natives of the White River bottomlands as Choctaw logs, after the tribe of Indians in this part of the country. After undergoing certain natural processing, Choctaw logs have a buoyant quality that makes them useful as floats for White River cabinboats. According to P. J. Van

Cypress Bayou, Tichnor, Arkansas County, Arkansas. Tree on left with flared base is tupelo gum; tree on right with buttressed base and knees is bald cypress. Cypress Bayou flows between the Arkansas and White Rivers.

Huizen, an authority on the natural history and folklore of this region, the process of Choctawing seems to occur only in the red oak species, particularly Nuttall oak. Old mature trees that fall on the ground attain a glazed appearance; the tree cells all become empty, but are sealed and look white and almost like empty honeycomb. These logs float for years.

The pecans are often difficult to tell apart. The bitter pecan or water hickory has the more flaky bark. The fruits of these two species are distinctly different, as are the sites on which they grow. Buckwheat vine, so named because it is in the buckwheat family (Polygonaceae), grows mainly along the edges of openings, often forming a thick hanging mat-like growth. The thick cover is a favorite winter roosting site for robins and rusty blackbirds.

This interesting portrait of a deer was made by P. J. Van Huizen in the White River bottoms of southeastern Arkansas.

Deer concentrated in uplands during overflows when the White River bottoms are flooded. (Photograph by Peter J. Van Huizen.)

I have been impressed by the size of the cypress "knees" that I have seen in parts of the White River bottom. Several were five feet high. This is higher than any that I have seen in the Southeastern swamps. Cypress "knees" are a characteristic feature of most Southern swamps and bottomlands. They usually surround the buttressed base of the parent cypress tree, rising several feet above water level.

However, in the bottomlands of the Cache River, a tributary of the White that flows through northeastern Arkansas, I found cypress knees ten and eleven feet in height. In thirty years of exploring the Southern swamps and bottomlands, I had previously not seen knees more than six feet tall, nor had my colleagues in the U.S. Fish and Wildlife Service who have spent a number of years at such places as Reelfoot Lake, Okefenokee, Santee River, and many other swamps. The height of cypress knees corresponds closely to the average high-water level for the locality. At the time of my visit to the Cache River slough, the water was about five feet in depth.

The function of cypress knees does not seem to be known, although it has been hypothesized that knees are aerating organs through which gas exchange occurs with submerged roots. However, experiments by Kramer and others (1952) seem to indicate that cypress knees do not play an important role as aerating organs. In fact, it appears that cypress trees grow as well without knees as with them.

Below: Mature bottomland forest of Nuttall oak, sweet pecan, and sweet gum. White River bottoms near St. Charles, Arkansas. Circa 1956. This park-like stand is prime wild turkey foraging habitat. (Photograph by Peter J. Van Huizen.)

Above: Trees of the primeval forest of the Southern bottomlands looked like this overcup oak in the White River bottoms of southeastern Arkansas. Such giants as exist today are mainly culls. There was a bear's den in one of the hollow upper limbs of this tree when the photograph was made in the early 1950s. Lloyd MacAdams, the woodsman standing next to the tree, is a leading authority on the lower White River bottoms. (Photo by P. J. Van Huizen.)

The old river town of Clarendon, Arkansas, on the lower White River, winter 1968.

The cabin boat is the permanent home of the people of the lower White River country who make their living by commercial fishing, digging clams, and trapping. (Photograph by P. J. Van Huizen.)

Shaggy or flaky bark is characteristic of the bitter pecan or water hickory that grows in the White River bottoms of southeastern Arkansas. (Photograph by P. J. Van Huizen.)

In my opinion, one of the most valuable wildlife resources of this wilderness is the native turkey population. This is one of the few places where the pure strain Eastern race of the wild turkey can still be found.

During the period 1950–55, I obtained considerable information on foods of the wild turkey in this area, with the help of Pete Van Huizen. Crops and gizzards were obtained from hunters during the spring gobbler hunting season, and droppings were gathered at other seasons from beneath roosts and along woodland trails. Most of the crops and gizzards were from birds taken on Big Island at the mouth of the White and Arkansas Rivers.

In this area, fruits, seeds, and herbaceous leaves form the great bulk of the turkey's food. Insects, although they occur frequently in crops, gizzards, and droppings, consistently compose little of the total volume.

Crops of turkeys obtained during the spring gobbler hunting season (early April), following an autumn when there was a good crop of sweet pecans, usually contained from two to fifteen whole nuts. During years of low pecan production, crops and gizzards were usually crammed full of sugarberries, poison ivy berries (one crop contained 3,000 poison ivy seeds), acorns, and flowers. Flowers most often eaten were crossvine, buttercup, and catkins of cottonwood poplar and oak.

In summer, when turkeys feed mostly in dry lake beds and on sand bars, their droppings indicated that seeds of crab grass and sprangletop grass, fruits of the snowbell, and grasshoppers, crickets, and scarab beetles were important foods.

Turkeys worked on first bottom ridges a great deal in fall and winter, feeding mostly on nuts and fruits. Acorns, sweet pecan nuts, sugarberries, grapes, fruits of possumhaw and black gum, and blades of grass were the usual foods consumed. Two insects, the Southern green stinkbug (*Nezara viridula*) and the wheel bug (*Arilus cristata*), also were taken frequently. Both insects occur commonly

Bald cypress knees ten feet tall, Cache River bottoms, northeastern Arkansas, February 1967. Eugene Hester, the man in the picture, is an even six feet tall, and is standing on a part of a cypress knee.

throughout the winter beneath the leaf mantle of riverbottom hardwoods. The diet of turkeys in the White River bottoms also included such items as jack-in-the-pulpit leaves, Solomon's seal berries, the fruits of supplejack or rattanvine, and pods of vetch.

Lloyd MacAdams, who has had a lifetime of experience in the White River bottoms, told me that among the favorite foods of bears are the acorns of overcup oak, pokeberries, persimmons, and black gum berries. Mac tells me that he has seen bears up in overcup oaks eating acorns which had not yet fallen, while ignoring the acorns of Nuttall oak which were lying plentifully on the ground nearby. White River fox squirrels seem to have an opposite preference. In the spring when foods are in shorter supply, bears may even feed on the roots or shoots of the bitter pecan. Mac knew of a bear den in a hollow limb of a hugh overcup oak that was in use almost every year for twenty-five years.

Since the White River wilderness is strategically located in the bottleneck of the Mississippi Flyway, mallard ducks by the thousands are funneled into that area. An estimated 500,000 ducks winter in White River country, and mallards comprise 90 percent of the population.

The mallard population reaches a peak in late winter when the bottoms are flooded, making acorn mast available. Acorns of the willow and water oaks are utilized most by the wild ducks. In a study of acorn mast production in the White River bottoms, Cypert and Webster found that a 124-year-old water oak with a height of ninety-eight feet produced 28,360 acorns in one year.[1]

The ducks compete with fox squirrels, wild turkeys, bronzed grackles, blue jays, woodpeckers, and small rodents for the acorns. Cypert and Webster found

Winter concentration of mallards, Moon Lake, White River bottoms, Desha County, Arkansas. Circa 1950. (Photograph by Peter J. Van Huizen.)

Hooded merganser flies from nest hole, White River bottoms, southeastern Arkansas. This is south of the main breeding range. (Photograph by P. J. Van Huizen.)

that blue jays and red-headed woodpeckers removed as much as 12.7 percent of the acorn crop from trees in their study area.

Great flocks of migrant bronzed grackles, sometimes numbering in the millions, come ahead of the ducks when the bottoms are still dry and clean up some of the acorn mast in the fall. The manner in which the great flocks roll through the bottoms is reminiscent of the days and ways of the passenger pigeon. They filter through the woods at all levels, picking off acorns and gum berries from high in the trees on down to the ground. Few leaves are left unturned as they work along the ground searching for fallen poison ivy berries, insects, amphibians, and snails. The leaf mantle often is so stirred up by the grackles that it looks as if a flock of wild turkeys had been feeding there.

One winter Peter Van Huizen showed me a grackle roost in a canebrake near St. Charles, on the edge of the bottomlands. Earlier in the fall the roost had been occupied by thousands of robins. The ground beneath the roost was covered with about six inches of guano, part of which was an accumulation of previous years. The guano was covered with sugarberry seeds deposited by the robins and with pellets of discarded rice hulls regurgitated by the grackles. Rice left from the fall harvest was gleaned by the grackles from stubble fields on the nearby Arkansas Grand Prairie.

The rusty blackbird is sometimes an associate of the bronzed grackle in the bottomlands. No wintering bird from the North is more closely associated with the White River lowland forest than the rusty. This blackbird likes to get its feet wet and does most of its foraging in the the swampy parts of the bottoms. It walks about flipping over wet leaves in search of snails and aquatic insects. It also feeds on vegetable matter, and stomachs that I examined contained fruits of sugarberry, grape, and acorn mast. Sometimes the rusty steals bits of acorn from grackles.

In the eastern Arkansas hardwood bottoms, willow oak mast is one of the more important foods of the grackle. The grackle places an acorn crosswise in its bill, and with the aid of a sharp keel-like appendage on the inside of the upper part of the bill is able to cut the acorn husk in two as it revolves the nut around in its bill, applying pressure at the same time. A rusty, unable to open or husk an acorn without considerable effort, remains close to a grackle until the husk has been removed, then runs up to the grackle which drops the acorn to make a pass at the smaller blackbird. Before the grackle can recover the nut, the more agile rusty has made away with it.

By 1970, the lower White River country was still an extensive forested area with many big trees, about 100 black bears, an estimated 1,000 wild turkeys, alligator gar in the river measuring up to 8 feet and weighing up to 230 pounds, and alligator snapping turtles that weighed as much as 150 pounds.

Gleaners of the bottomlands. Common grackles filter through the bottomlands at all levels in the forest foraging for acorn and beech mast, black gum berries; and turning over the leaf litter for lizards, insects, and snails. These great flocks of grackles, sometimes numbering 10,000 birds, are the nearest thing to the passenger pigeon in their method of foraging. (Photograph by Frederick C. Schmid.)

I'ON SWAMP

In the days when many new species and geographic races of North American birds were being discovered, I'On Swamp, located in the heart of the South Carolina Low Country, was the collecting ground of two of the South's pioneer ornithologists. The Rev. John Bachman, who for more than fifty years was the pastor of St. John's Lutheran Church in Charleston, collected the first specimen known to science of Bachman's warbler in this swamp. The new warbler and other new bird discoveries of Bachman's were given to his good friend John James Audubon, who described these birds in his *Ornithological Biography*. Bachman's warbler, America's rarest songbird, was discovered in 1833.

This new bird was thereafter unobserved in South Carolina until 1901, when it was "rediscovered" in the Low Country by Arthur T. Wayne. Wayne conducted most of his studies of Bachman's warbler in I'On Swamp, where he located nine of the only sixteen nests of this species that have ever, and probably ever will be, found, as the bird is near extinction.

I'On Swamp is named for Colonel Jacob B. I'On, who owned the land in the early 1900s. The U.S. Geological Survey quadrangle maps have mistakenly labeled the area Iron Swamp. The map maker either mistook the apostrophe in this unusual name for an *r* or a Northerner doing the field mapping thought some Southerner with a Charleston accent was saying *iron* when he said *I'On*.

I'On Swamp is now a part of the Francis Marion National Forest, which lies along the coast between Charleston and the Santee River. The swamp is about twenty miles northeast of Charleston near the village of Awendaw; and is just west of U.S. Highway 17. Iron (I'On) Swamp Road, leading from highway 17, is difficult to locate without a detailed map of the Francis Marion Forest area (U.S. Geological Survey, Sewee Quadrangle). However, one can inquire locally regarding access to the swamp.

An interesting feature of the I'On Swamp area is the pineland gall or bay gall—a long, narrow swamp that extends in a straight line for several miles through the piney woods. It rarely exceeds a few rods in width. Bays and gums are usually the predominant trees of the pineland galls. The term gall refers to the gallberry bushes that occur in this type of swamp; but it refers also to the unattractiveness of such an area.

It was in the tallest trees in I'On Swamp that I saw my first Wayne's warbler. The breeding range of this Southeastern race of the black-throated green warbler is a narrow strip of Outer Coastal Plain extending from the Dismal Swamp region of Virginia to about Charleston, South Carolina (see American Ornithologists' Union Check-list, 1957).

The Bachman's warbler, rarest songbird in North America, is close to extinction. A single bird is seen about every other year. (Illustration by John W. Taylor.)

From the time that it became recognized as a subspecies of the much more widely distributed Northern form, the Wayne's warbler has been mainly identified with the South Carolina Low Country, where it was discovered. Sprunt and Chamberlain tell about its discovery in *South Carolina Birdlife* (1949):

That South Carolina has a very interesting ornithological history is beyond question, and not the least striking evidence of the fact is this dainty, fragile bird of the cypress lagoons. . . . It properly bears the name of its discoverer.

For many years Arthur T. Wayne was puzzled by the fact that he saw and heard Black-throated Green Warblers on the coast from late March to June but could find no evidence of nesting. Then on April 11, 1917, he saw a female carrying nesting material. He made a determined effort to resolve the mystery and, in the Spring of 1918, took specimens, compared them carefully, and found constant variation between these coastal birds and typical Black-throated Greens. The specimens were sent to Outram Bangs at the Museum of Comparative Zoology in Cambridge, Massachusetts. From them, Bangs described a new race, naming it Dendroica virens waynei (*Proc. New Eng. Zool. Club*, 6:1918, 94). Thus Wayne's Warbler came into scientific being. The principal subspecific difference is the much smaller and more delicate bill, a character which is at once apparent in specimens in the hand.[1]

68

The Rev. John Bachman. He collected the type specimen of the Swainson's warbler along the banks of the Edisto River in South Carolina in 1832. (Photograph courtesy of E. Milby Burton of the Charleston, S.C., Museum.)

In April 1958, Robert E. Stewart and I spent several fruitless days in I'On Swamp looking for Bachman's warbler. Most of the recent sightings of this very rare bird have been in this section of the South Carolina Low Country where it was discovered.

About a month after our search for this elusive bird in I'On Swamp, I saw my first Bachman's, a singing male, about fifteen miles south of Washington, D.C., on the Virginia side of the Potomac. Dr. Irston Barnes led me to its "staked out" territory in a floodplain forest along Pohick Creek. This same bird, presumably, had been discovered in the same place four years earlier in 1954, by Harriet Sutton and Morgan Gilbert. During the 1954 sighting some 100 bird watchers saw the bird over a period of several weeks, but in 1958 when I saw it, only five persons were so privileged.

I was back in I'On Swamp looking for the rare warbler in 1963. About three miles east of the swamp, along the Bull's Island Road, Rhett Chamberlain of Charleston showed me my second and only other Bachman's.

Bachman's warbler is so rare today that several years may go by without any being seen. The fact that there is an increasing multitude of bird watchers and that they are seeing fewer and fewer Bachman's warblers is further evidence of its rarity. Also, none have been reported to have struck TV towers along former major migration routes of this species. The Tall Timbers TV tower a few miles north of Tallahassee, Florida, lies in the path formerly followed by this warbler in its spring migration through northern Florida. During the eleven-year period, 1955–66, 30,000 birds of 170 species have been picked up at the base of this tower, but not a single Bachman's was among them.

All Bachman's warblers seen in recent years seem to have been "stumbled upon." Two of my ornithological friends, Tom Burleigh and George Sciple, each collected a specimen on islands off the coast of Mississippi in the 1940s. Burleigh was squatting in the bushes when he heard one and soon collected it. Sciple was collecting migrating warblers in connection with an encephalomyelitis study for the Georgia Public Health Service when he fired into a flock of small birds, and among the several specimens to drop out was a Bachman's. In another unusual situation, some bird-watchers at Charleston, South Carolina, were attracted to a bird attempting to extricate itself from a spiderweb. As they approached the bird, it freed itself, and was never identified. After lowering their binoculars, they heard a strange song directly overhead. On looking up they saw a Bachman's warbler. Almost as surprising as its appearance was the presence in a dry pine woods of this classic inhabitant of swamplands.

There seems to be no reasonable explanation for the rarity of this tiny songbird. It is not difficult to understand, however, why several larger species have become rare or extinct because of illegal hunting.

The gradual disappearance of Bachman's warblers has been linked with the cutting over of the virgin swamp and bottomland forests in the early part of the twentieth century. L.S. Golsen reported that they had usually been abundant in Bear Swamp near Autaugaville, Alabama, until 1928, when, possibly due to the cutting of the timber, the birds disappeared.[2]

Possibly the disappearance of the canebrakes that formed the understory in many of the primitive bottomland and swamp forests inhabited by this warbler may be one factor related to its extirpation in some areas. During the 1940s and 1950s, I visited the three areas where Bachman's warbler was formerly known to be a common breeding bird (Bear Swamp, Alabama; Sunken Lands along the St. Francis River in southeastern Missouri; and I'On Swamp, South Carolina). Cane used to be prominent in the understory in all three places. Parts of these areas still had some fine forests but a paucity of cane understory.

While Bachman's is said to be generally high-ranging like the yellow-throated warbler, the two males that I studied on their breeding territories usually sang and foraged below 50 feet in trees that were 70 to 100 feet in height. Often they fed within five and ten feet of the ground. Stomachs of birds collected fifty or more years ago contained some of the same kinds of food upon which recent Bachman's have been seen feeding.

Several ornithologists have suggested that some birds may perish in hurricanes during their southward migration across the Florida Straits enroute to the wintering home in Cuba. The Bachman's warbler is one of the earliest migrants, some birds reaching the Florida Keys in late July and early August, and most of them migrating across the Straits in August. Their migration coincides quite closely with the main hurricane season.

Apparently the Bachman's warbler is a relict species. Amadon regards a relict as one "whose numbers or range or both have undergone drastic reductions." They "tend to become more and more restricted both geographically and ecologically because they are unable to compete successfully with other species."[3] Thus they would be expected to occur only in the optimum habitat. Interestingly, the breeding territories occupied by four males in recent years were in atypical habitat, and two of the males occurred in areas that probably were beyond the normal breeding range. This suggests that traditional breeding grounds in some areas have been altered.

Further, according to Amadon, migratory relict species "frequently have winter ranges that are as circumscribed as the breeding range itself. This suggests that the populations of such species may be, in some cases, more rigidly controlled by conditions on the wintering (or migratory) areas than by those encountered during the nesting season. . . . Further, the winter ranges of relict species are often insular as is the case of the Bachman's Warbler, presumably because competition is thereby reduced." Bachman's warbler is known to winter only in Cuba and the nearby Isle of Pines. Native habitat of all types has been drastically reduced or altered on these islands in the past several decades.

THE PINETOWN POCOSIN

IF a nearly impenetrable thicket exists in the South it is the evergreen shrub-bog community. This is the "pocosin" of the Indian, and "bay" or "Carolina bay" of the early settlers. It is a sort of upland bog.

Pocosins are poorly-drained flatlands or savannas of interstream areas, and are slightly higher than surrounding swamps and river bottomlands. For this reason they were given the name *pocosin* or "swamp-on-a-hill" by the Indians (although they are hardly hills as we know them). Apparently they were called Carolina bays because of the several kinds of bay trees in this plant community. Some pocosins or bays occur as shallow, elliptical depressions and are thought to be of meteorite origin. Pocosins reach their best expression in the Outer Coastal Plain of North Carolina, and occur also in parts of South Carolina, Georgia, and extreme southeastern Virginia.

Pocosins are a combination of water-logged peaty soil, a ground cover of sphagnum moss in wetter sections, dense broad-leaved shrubs, small trees that

Pocosin or Carolina bay, Pinetown, Beaufort County, North Carolina, January 1960. Vegetation mainly pond pine and ericaceous shrubs.

Gallberry or inkberry, common
shrub of pocosins and moist
coastal pine flatwoods.

are mostly evergreen, thorny vines known as greenbrier, and a pond pine overstory. They have virtually the same appearance the year round.

The water table fluctuates between a high near ground level to a low many feet below. The periods of relative dryness eliminate any invading marsh plants, while the bog shrubs with their pointed, leathery, wax-coated leaves are adapted to retain their moisture. The low oxygen content of bog soils during the high-water periods also discourages non-bog species. Microscopic forms—bacteria, fungi, etc. —are destroyed during high water levels in the surface soil layer, resulting in the prevention of decay, and accumulation of leaves and other bits and pieces which are preserved and become peat. Nitrates are virtually absent because nitrogen-fixing bacteria do not survive in these soils.

Common shrubs of pocosins are fetterbush or hurrah-bush, ti-ti, honeycup, inkberry or gallberry, shining inkberry, sweet pepperbush, Virginia willow, leucothoe, chokeberry, blueberry, and huckleberry. The leaves of these species of shrubs are elliptical in shape and coriaceous or leathery in texture. Shrub bog plants are adapted to long periods of water logging and drought, and to rapid regeneration following fire.

Partly opened flower of Gordonia or loblolly bay.

Prolonged drought or drainage of pocosins usually results in a succession from broad-leaved evergreen shrubs to bay trees, i.e., sweet bay, red bay, loblolly bay, and then to extensive pure stands of switch cane.

I "discovered" pocosins while scouting for blackbird roosts in the Outer Coastal Plain of eastern North Carolina. The first of several large roosts investigated was located a few miles northwest of Pinetown, Beaufort County. The Pinetown roost contained several million blackbirds and a few thousand robins. So dense was this briery jungle that it often took an hour of hacking away with a machete to move forward a hundred yards.

Black bears were fairly common in the Pinetown pocosin at the time of my visits in the winters of 1958 and 1959. An examination of their scats or droppings indicated a fondness for the fruit of the inkberry or gallberry. Where cornfields bordered the pocosins I found corn stalks that had been carried two or three hundred feet into the woods by bears.

Also in the late 1950s, the red-cockaded woodpecker, an endangered species, was one of the common breeding birds of the Pinetown pocosin. It was as numerous there as at Thomasville, Georgia, which probably has the best stands of virgin longleaf pine still in existence.

The red-cockade nests only in mature or overripe pines. Nest cavities are excavated in heartwood which has been softened by redheart, a disease caused by the fungus *Fomes pini*. Modern forestry practices eliminate most of these trees. Trees are cut on a 25–35 year cycle, before they reach old age when they could be used by nesting red-cockaded woodpeckers.

Red-cockades are usually found in small groups of several individuals. Some ornithologists think that these are family groups. For a number of evenings I watched a family group of five birds go to roost in the same dead pine. There were five holes in the tree, one above the other, and each bird selected a different hole.

Since there seems to be very little information in popular literature on the subject of pocosins, I would recommend for further reading two publications by B. W. Wells: *The Natural Gardens of North Carolina*, published in 1967 by the University of North Carolina Press, Chapel Hill; and *Vegetation of Holly Shelter Wildlife Management Area*, published in 1946 by the North Carolina Department of Conservation and Development, Division of Game and Inland Fisheries, Raleigh.

Bear tracks on trail through pocosin. 25-cent piece lies beside tracks.
Pinetown, Beaufort County, North Carolina, January 1960.

Southern magnolia and spruce pine, indicator plants of Dudley's Hammock near Valdosta in southern Georgia. This hammock is surrounded by swamp black gum-pond cypress swamp and low flatwoods of slash, longleaf, and pond pines. Other interesting plants of Dudley's Hammock include dahoon (an ilex), green-fly orchid, and devilwood or wild olive. January 1967.

DUDLEY'S HAMMOCK

About forty miles west of Okefenokee lies a smaller but similar swamp known as Grand Bay. Its approximately 20,000 acres are in Lowndes and Lanier Counties, between the towns of Valdosta and Lakeland, Georgia. Like Okefenokee, Grand Bay has its cypress bays, cypress heads, maidencane prairies, lakes and ponds, savannahs, and hammocks. In the early part of this century, people lived in several of these hammocks or, as they are sometimes called, islands.

Because it is slightly more elevated than the surrounding swamp or flatwoods, a hammock has a different appearance and is relatively dry. As in pocosins, the trees, shrubs, and vines are mostly evergreen. However, a hammock is characterized more by broad-leaved evergreen trees, a pocosin by broad-leaved evergreen shrubs.

One such hammock, in which I collected plant specimens in December 1967, is known as Dudley's Hammock. This hammock is surrounded mostly by a swamp black gum-pond cypress swamp and partially by low wet flatwoods composed of slash, longleaf, and pond pines. The soil of the hammock is similar to that

Resurrection fern or tree polypody growing on limb of live oak.

of the flatwoods, rather sandy. The saw palmetto in the hammock is shared with the flatwoods, but otherwise the hammock has its own unique features.

The most conspicuous indicator plants of Dudley's Hammock are Southern magnolia and the spruce pine. In its natural environment such as is this area, Southern magnolia grows tall and straight, unlike specimens which broaden out when grown in the open. The spruce pine has a rather restricted range, occurring only in coastal South Carolina, the southern parts of Georgia, Alabama, and Mississippi, southeastern Louisiana, and northern Florida.

Other major overstory components of the hammock forest are live oak and

The green-fly orchid is the northernmost of the tree orchids. In Dudley's Hammock it grows only on mature Southern magnolia trees. December 1967.

water oaks, loblolly pine, pignut hickory, and sweet gum. The more important shrubs are staggerbush, farkleberry, and Elliott's blueberry. The number of vines is surprisingly small and includes mainly yellow jessamine and catbrier.

I observed only two small ground plants, partridgeberry and dwarf smilax. Unlike other smilaxes, the dwarf smilax in this hammock grows as a small upright stem of about six inches in height, rather than as a vine, and it usually has only one or two leaves. Two ilexes, American holly and dahoon, were present.

The only fern present was the tree polypody or resurrection fern, which grows on live oak and Southern magnolia. The green-fly orchid, northernmost of the tree orchids, grew only on mature magnolia trees. Spanish moss festooned most of the trees.

The mountain vireo and the orange-crowned warbler were two interesting wintering birds that I observed in the greenery of the hammock.

TREES, SHRUBS, AND VINES
OF A 2-ACRE PLOT IN DUDLEY'S HAMMOCK

Trees	*Percent Composition*
Southern Magnolia	28
Water Oak	23
Spruce Pine	18
Live Oak	18
Loblolly Pine	5
Pignut Hickory	5
Sweet Gum	3
Swamp Chestnut Oak	trace
Horse-sugar	trace

Shrubs or Shrub-like Plants	*Percent Composition*
Staggerbush	35
Farkleberry	25
Elliott's Blueberry	20
American Holly	10
Saw Palmetto	10
Devilwood	trace
Dahoon	trace
Myrtle	trace
Downy Leucothoe	trace

Vines	
Yellow Jessamine	80
Catbrier	15
Sawbrier	5

THE EDISTO AND BRIER CREEK

Dᴜʀɪɴɢ the thirty years that I roamed the Southern swamps from the Pocomoke on the "Eastern Shore" of Maryland to the Atchafalaya in southern Louisiana, my principal objective usually was the little known Swainson's warbler. By 1971, I had written a monograph on this phantom of Southern swamps and river bottomlands.

It was inevitable that sooner or later my path would lead to the two swamps where the Swainson's was discovered. One of these is Big Swamp along the Edisto River near the present village of Jacksonboro, South Carolina; the other, Brier Creek Swamp, a tributary of the Savannah River, located near the extinct town of Jacksonboro, Screven County, Georgia.

South Carolina's Jacksonboro and nearby Parker's Ferry Landing, located about twenty-five miles south of Charleston, were collecting localities of the Rev. John Bachman. This area was the type locality of two of Bachman's discoveries, the Swainson's warbler, and the Bachman's sparrow. The sparrow is not a swamp bird, but is found in the nearby piney woods.

Bachman collected several specimens of the warbler along the Edisto in 1832. He presented the unknown bird to his close friend Audubon, who described it, naming it for the English ornithologist, William Swainson.

Audubon's painting of the Swainson's warbler appeared in his *Birds of America*[1] and the description of it in his *Ornithological Biography*[2]. The type specimen was given to the U.S. National Museum by Spencer F. Baird, one-time secretary of the Smithsonian Institution, who acquired it from Audubon.

The discovery of this new species by Bachman was described as follows:

I was first attracted by the novelty of its notes, four or five in number, repeated at intervals of five or six minutes apart. These notes were loud, clear, and more like a whistle than a song. They resembled the sounds of some extraordinary ventriloquist in such a degree, that I supposed the bird much farther from me than it really was; for after some trouble caused by these fictitious notes, I perceived it near me and soon shot it.[3]

While John Bachman gets the credit for the discovery of the Swainson's warbler, John Abbot, a Georgia naturalist, had apparently collected a specimen some twenty-five years earlier but made no public record of the event. However, he made an identifiable portrait of the bird. Many of Abbot's Georgia bird paintings were deposited in the British Museum and the Boston Society of Natural History. Those paintings, including that of the Swainson's warbler, deposited in Boston, now repose in the Harvard College Library. Walter Faxon, one of the first persons to study Abbot's paintings, made the following remarks about the painting of the Swainson's warbler:

80

John Abbot's painting of the Swainson's warbler, which he called the "Swamp Worm-eater." (Illustration by permission of the Harvard College Library.)

On looking through the Abbot bird-portraits several arrest the eye from their historic interest. Plate 68 is a good representation of Swainson's Warbler, drawn at least a quarter of a century before this species was described and named by Audubon. On the reverse of the plate is the following autograph note by Abbot: L. 6. May 8. Swamp—Swamp Worm-eater.[4]

After Bachman collected his historic five specimens in 1832, the Swainson's warbler was almost a lost species for the next fifty years. According to William Brewster, only eight or nine specimens were collected during that period.[5] Then in 1884, Brewster and Arthur T. Wayne made significant collections and studies in the vicinity of Charleston, South Carolina.

Wayne reported the first nest and eggs of the Swainson's warbler known to

81

science. They were found near Charleston, South Carolina, on June 6, 1885.[6] Troup D. Perry of Savannah, Georgia, found a nest twenty-two days earlier (May 16) but did not report his discovery as soon as did Wayne.

The swamps along the Edisto and Savannah, where Bachman and Abbot collected the first specimens of the Swainson's warbler in the early 1800s, were still good Swainson's warbler habitats in the 1970s. Some of these little known warblers continue to pass through during spring and fall migration, and a pair occasionally nests there.

Since the Swainson's warbler was thought to be restricted to the Coastal Plain or Low Country near the coast, ornithologists were surprised to learn by the 1930s that this little known warbler was a locally common breeding bird to an elevation of about 3,000 feet in the Southern Appalachians. Before the 1930s there had been several records from the Piedmont suggesting the possibility of an up-country population. While there are a few records of its occurrence in the Piedmont Province during the breeding season, there appear to be no breeding concentrations in this in-between area. The swamps and floodplain forests of the Coastal Plain, and sections of the mixed mesophytic forest where this species occurs in the mountains, are more humid than most of the forests of the Piedmont.

Breeding birds from the Southern Appalachians differ from the Coastal Plain birds in that the underparts tend to be whiter (less tinged with yellow). On the basis of the plumage variation in these two geographically separated breeding populations, the mountain bird was described as a new subspecies by Gorman M. Bond and me, when we were in the Division of Birds at the U.S. National Museum. This new geographic race or subspecies is known as *Limnothlypis swainsonii alta* (Appalachian Swainson's Warbler).

Type locality of the Swainson's warbler, Edisto River near Jacksonboro, South Carolina. Photograph made in January 1958.

REELFOOT

REELFOOT is best known as the lake that was spawned by an earthquake. The New Madrid earthquake, identified by the name of a nearby Missouri town, occurred in the winter of 1811–12, in the Mississippi Valley section of northwestern Tennessee, southeastern Kentucky, southeastern Missouri, and northeastern Arkansas. Geologists have reported it to be the most violent earthquake in the history of the nation. Some of the pioneer settlers said that on the day of the first shock, December 15, 1811, the Mississippi ran backwards for several minutes! After the earthquake, the area became known as the "Sunken Lands" and to this day, the St. Francis River basin of the Missouri "Boot-heel" still goes by that name.

Reelfoot was named for a Chickasaw Indian chief with a deformed foot, who lived in the northwest corner of Tennessee when the present lake was a huge cypress swamp. Reelfoot Lake was formed when the cypress swamp sank during the earthquake. During periods of low water, some of the stumps of the ancient forest extend above the lake's surface. The stumps of the drowned or submerged cypress make boat travel on the lake slow and hazardous. In order to adjust to these conditions, Reelfoot boats and oars were developed. The oars are jointed, and the oarsman faces forward when rowing.

Most of Reelfoot Lake is in Lake and Obion counties, Tennessee, with the north end extending just beyond the border into Fulton County, Kentucky. The lake lies in a general north-south direction and is about twelve miles long from a point near Tiptonville, Tennessee, to just over the Kentucky line. Its width varies from one to four miles.

Reelfoot Lake and its surrounding countryside is near the north end of the Mississippi River Delta or Alluvial Plain, but has the appearance and most of the characteristics of the deltalands down river in eastern Arkansas, western Mississippi, and eastern Louisiana. If Spanish moss occurred this far north, Reelfoot country would look just like the deep south delta. Spanish moss reaches its northern limit on the west side of the Mississippi River in the southeastern corner of Arkansas, and in Mississippi just below Greenville. This is about 200 miles south of Reelfoot. On the Atlantic Coast, Spanish moss reaches its northern limit at Norfolk, which is about the same latitude as Reelfoot.

In one respect, Reelfoot is like the Great Dismal Swamp of the Atlantic Coast, as the northern outpost of many Southern swamp plants and animals. The Fulton County, Kentucky, extension of Reelfoot is approaching the northern limit of the cottonmouth moccasin, the red-bellied water snake, mud snake, green water snake, swamp rabbit, rice rat, cotton rat, and the water turkey or anhinga. The

Swainson's warbler is locally common at Reelfoot, but beyond that point is quite sparsely scattered to extreme southern Illinois, really not far away. The alligator apparently never occurred this far north in the Mississippi Valley. At one time, its range extended to Memphis.

There are about 20,000 acres in the Reelfoot complex, which consists of approximately 7,000 acres of open water, 6,000 acres of swampy woods and stream bottoms, 3,000 acres of marsh, and additional acreage of mixed plant communities.

The greatest depth of the lake is about twenty feet, but the average depth is about five feet. Plant growth is correlated with water depth. My colleague John Steenis conducted aquatic plant management studies at Reelfoot for five years, and his observations of the succession from deeper to shallower waters and then to land can be summarized as follows:

1. Submerged aquatic growth in water up to ten feet in depth with coontail usually dominant. Coontail is a free-floating underwater plant.
2. Dense stands of lotus or "yonkapin" usually dominant in water three to eight feet in depth. Yonkapin has a huge leaf, some measuring thirty inches in diameter.
3. Spatterdock or "mulefoot" in water from 2 to 5½ feet.
4. Extensive stands of giant cut-grass, from a few inches to three feet. Giant cut-grass is a marsh type plant that in places forms extensive dense stands. It attains an average height of about six feet, and persists through the winter, having a blanched appearance. In the Southeast, the early settlers called it the "white marsh."
5. Willow and buttonbush in damp soil to watered areas a few inches in depth.
6. Bottomland trees including bald cypress, sweet gum, sugarberry, sycamore, pecan, cherrybark oak, American elm, overcup oak, willow oak, box elder, red maple, water locust, planer-tree, and other hardwoods in damp soils near the water.[1]

Because of its variety and interspersion of aquatic habitats, Reelfoot is a mecca for water birds. And perhaps the most spectacular of these are the egrets and herons, locally called cranes. They are colonial nesting birds, and during the breeding season form rookeries or heronries often with thousands of birds. During the 1920s and 30s, one such heronry at Reelfoot known as "Cranetown" was visited by hundreds of bird watchers. The tenants of this great heronry were American egrets, great blue herons, black-crowned night herons, double-crested cormorants, and water turkeys or anhingas. Cranetown survived for a number of years because of its relative inaccessibility. Most of the nesting was in huge cypress trees, some of which reached heights of 125 feet. Eva O. Gersbacher

Opposite: A cypress break with pad plants (waterlilies) at Reelfoot Lake, northwestern Tennessee. Principal pad-leaf species at Reelfoot are white waterlily or "bonnets," lotus or "yonkapin," spatterdock or "mulefoot," and watershield or "dollarpad." (Photograph by John Steenis.)

Cypress stumps and spatterdock or "mulefoot" (a species of waterlily) in Reelfoot Lake, Tennessee. Reelfoot Lake was formed when a cypress swamp sank during the New Madrid earthquake of the winter of 1911–12. (Photograph by John Steenis.)

The stately American or common egret is one of several large wading birds seen at Reelfoot Lake. Their numbers appear to be diminishing in this area and in other sections of the Lower Mississippi Valley, possibly due to the widespread use of pesticides during the past twenty years. (Photograph by P. J. Van Huizen.)

made a rather intensive study of this colony in the summer of 1938, and estimated the number of nesting species as follows: 3,500 American egrets, 1,000 double-crested cormorants, 350 great blue herons, 225 black-crowned night herons, and 200 water turkeys or anhingas. Cranetown is no longer in existence, but smaller rookeries or heronries are presently found in the region.

An extremely interesting adjunct to Cranetown was the pair of nesting peregrine falcons that had their eyrie in the hollow of a big cypress tree near the heronry. Peregrine falcons are usually thought of as being cliff nesting birds in such places as the Appalachians; such hollow tree nesting is an extreme rarity. The nesting at Cranetown was first reported in 1932, by Albert Ganier, a Tennessee ornithologist.[2] Some ten years later, Walter Spofford, a falcon expert, studied a nest in the same locality and photographed the young in their hollow-tree home.[3]

The prothonotary warbler is the most striking and one of the most abundant of the summer resident songbirds about the lake. These little yellow birds nest in woodpecker holes and natural cavities of cypress, willow, and other dead trees.

Today Reelfoot Lake is an important link in the Mississippi Flyway's chain of migratory bird sanctuaries. The U.S. Fish and Wildlife Service manages the upper one-third of the lake as a wildlife refuge. The Tennessee Game and Fish Commission manages the lower two-thirds of Reelfoot for public hunting and fishing.

The cottonmouth moccasin is the poisonous water snake of Southern swamps, river bottomlands, and rice fields. My colleague John Steenis saw sixteen in one day in a swampy section of Reelfoot Lake, Tennessee. When I worked in the Arkansas rice fields, I saw at least one a day during the summer.

In winter, the Refuge is the feeding and resting place of large aggregations of waterfowl, often numbering 100,000 birds. Mallards, pintails, gadwalls, and ring-necked ducks usually are the most numerous. Reelfoot is one of the few lakes in the central part of the country where a sizeable population of ringnecks occurs during the winter. The wood duck population has increased in recent years, and many nest there. The hooded merganser is a rare nesting bird at the lake that sometimes utilizes old pileated woodpecker nest holes. The wintering Canada goose population numbers up to 10,000 birds.

Reelfoot is an excellent waterfowl area because of the variety of habitat and food. Among the most important waterfowl foods are the pondweeds, submerged aquatic plants; smartweeds, emergent aquatic plants; duckweeds, smallest of the flowering plants; and acorns. The late winter and early spring rains flood the "pin" oak flats that border the lake, making the acorn mast available to the ducks. Seeds of spatterdock or yellow waterlily, locally known as "mulefoot," is a favorite food of ring-necked ducks.

It is in the swamp forests and drier hardwood bottoms around the edge of Reelfoot that the three poisonous snakes of the region are generally found. These are the copperhead, canebrake rattler, and cottonmouth moccasin. John Steenis counted sixteen cottonmouths in one day in the Bayou Du Chien section. I have seen more cottonmouths in this upper section of the Mississippi Alluvial Plain, including the Arkansas side of the river, than in any of the Southeastern swamps. An experienced cottonmouth man can tell when one of these snakes is near by its odor.

In connection with the canebrake rattler it should be mentioned that the great canebrakes of the Lower Mississippi Valley once covered as much as 100 acres of ground in the higher bottomlands of the Reelfoot country. Since the canebrakes, with poles sometimes measuring thirty feet, were found on the bottomland ridges, they were grubbed out very early, as this was the best ground of the rich bottomland soils. Today, only occasional small patches of cane are found in this area.

Reelfoot is one of the outstanding fishing areas of the South. The chief game fish are the largemouth bass, bluegill, and crappie (pronounced "croppy"). The commercial fish are mainly buffalo, carp, and catfish.

Judging from the scientific literature, more technical studies of fauna and flora have been made at Reelfoot than in any other Southern swamp. For further reading I would suggest in particular *Reports of the Reelfoot Lake Biological Station*, published by the Tennessee Academy of Sciences; and the *Migrant*, journal of the Tennessee Ornithological Society.

THE TENSAS

THE Tensas (pronounced "Tensaw") River of northeastern Louisiana winds through a remnant of what was until recently the finest tract of bottomland hardwoods in the Lower Mississippi Valley. The heart of the Tensas country was a virgin forest of gums, oaks, pecans, and hackberry, known as the Singer Tract. This tract was well known in the 1930s and 40s as the last stronghold of the ivory-billed woodpecker. The Singer Tract was located near Tallulah, in Madison Parish.

I know of no swamp or bottomland forests in the southern part of the United States that as recently as thirty-five years ago contained as many rare animal forms. In addition to the ivory-billed woodpecker, the Bachman's warbler (our rarest songbird), the panther or cougar, which in the eastern part of the country is now found only in the Everglades-Big Cypress region of south Florida, and the red wolf occurred in this vast wilderness. By the 1970s, only one of the four, the red wolf, may still occur there.

The best description that I have read of this famed wilderness is contained in a 1942 report by James T. Tanner, who made the only definitive study of the ivory-billed woodpecker.

According to Tanner, the Singer Tract contained about 120 square miles of virgin forest in 1934; at that time there were about seven pairs of ivory-bills in the tract.[1]

In the Singer Tract, ivory-bills occurred mainly in the ridge bottoms, feeding principally in the sweet gum-Nuttall oak forest. The overcup oak-bitter pecan association of the lower bottoms and the cypress-tupelo sloughs apparently were rarely used as foraging sites.

The ivory-bill, America's largest woodpecker, was a bird of the virgin forests, and its disappearance coincided with the cutting of the big trees. It was on the over-ripe and dying giants of the forest that the ivory-bill fed. Beneath the bark of dead branches of trees that were still partly alive, and the trunks of trees that had been dead a year or two, lived the wood-boring beetle larvae that comprised the big woodpecker's principal food. Unlike most other woodpeckers that do their foraging by digging or boring in dead wood, the ivory-bill did most of its searching for grubs by scaling off the dead bark from branches and trunks.

A pair of ivory-bills that Arthur A. Allen observed in Florida in 1924 did some of their feeding in dead pines, scaling off the bark of small and medium-sized pines that had been killed by fire. And one of the pair sometimes got down on the ground like a flicker, foraging around the fire-scarred trunks of saw palmettos.

In late summer and fall when wild fruits are plentiful, vegetable matter probably

90

Ivory-billed woodpecker at nest tree in the Singer Tract, Tensas River bottoms, Madison Parish, Louisiana, April 1935. (Photograph by James T. Tanner from National Audubon Society.)

constituted a sizeable percentage of the ivory-bill's diet. Three stomachs in the U.S. Biological Survey collection contained evidence of feeding on poison ivy berries, magnolia seeds, and hickory and pecan nuts, in addition to beetle larvae.

The ivory-billed woodpecker apparently was extirpated in the Singer Tract and Tensas country by the early 1940s. Lowery in his book *Louisiana Birds* (1955) states that the last authenticated report in the Singer Tract was in 1943.

A few ivory-bills probably were present in the Appalachicola River bottoms of northwest Florida, and the Big Cypress Swamp in south Florida during the

1950s. Since that time there have been reports from the Santee River swamp in South Carolina and the Big Thicket in eastern Texas. Neither of these reports was well substantiated.

A recent report from south-central Florida seems credible, however. An article by H. Norton Agey and George M. Heinzmann in the magazine *Birding* states: "An Ivory-bill was heard and/or seen on eleven occasions from 1967 through 1969. From a blown-down roost or nest tree, we recovered an entire nest hole. Its measurements exactly fit those given by Tanner for the Ivory-bill. Also we recovered a feather which Dr. Alexander Wetmore of the Division of Birds, U.S. National Museum, has verified as the innermost secondary (white) of an Ivory-bill." [2]

While the cutting of the virgin forests was probably the most important factor in reducing ivory-billed woodpeckers to the verge of extinction, certainly the killing of these birds by the Eastern woodland Indians, hunters, and museum collectors took a substantial toll. Mark Catesby, one of our earliest pioneer naturalists, wrote in his *Natural History of Carolina, Florida and the Bahama Islands* of the use of the bills of these great woodpeckers by the Indians: "The bills of these Birds are much valued by the *Canadian Indians,* who made Coronets of 'em for their Princes and great warriors, by fixing them round a Wreath, with the points outward. The Northern Indians having none of these Birds in their cold country, purchase them of the *Southern People* at the price of two, and sometimes three, Buck-skins a Bill." [3]

The late 1800s and early 1900s was a period when museums in this country were building up their scientific study skin collections. At this time the ivory-billed woodpecker was already known to ornithologists as an uncommon bird. As a result collectors were dispatched to Florida and Louisiana, and perhaps several other areas, to obtain specimens. The greatest number were taken from swamps in northern Florida. One collector procured some 200 birds in that region.

The marked increase in the number of hunters during the first half of the twentieth century helped to speed up the diminishment of the ivory-bill population. Both large woodpeckers, the ivory-bill and the pileated, have been shot for food and for curiosity.

The Tensas River country is the home of one of the three subspecies or geographic races of the red wolf, a Southern animal. The local form is known as the Mississippi Valley red wolf (*Canis niger gregoryi*). The type locality (that is, the area from which the original specimen was described) is Mack's Bayou, three miles east of the Tensas River; eighteen miles southwest of Tallulah, Madison Parish, Louisiana. The type specimen was collected by Ben V. Lilly on April 25, 1905. This specimen is in the U.S. National Museum. The race *gregoryi* was named in honor of Tappan Gregory, a mammalogist from the Chicago Academy of Sciences who photographed the black phase of the red wolf at night

The red wolf of the Lower Mississippi Valley. A few red wolves were still present in the Tensas River country of northeastern Louisiana in the 1950s. (Illustration courtesy of John W. Taylor.)

with a series of flash cameras, in the Tensas River wilderness in 1934, and wrote an interesting account of the experience entitled *The Black Wolf of the Tensas*.

A few cougars or panthers were present in the Tensas bottoms as late as the early 1940s, according to a report by Lowery published in 1943. But when I began exploring these bottoms in the early 1950s, most of the rare and spectacular species had been extirpated. I saw a few bear tracks, a lot of turkey sign and an occasional flock, and that will-o-the-wisp of the canebrakes, the Swainson's warbler.

The author banding blackbirds in an Arkansas blackbird roost. As many as
1,000 birds have been picked off branches by hand and banded in a single night.
(Photograph by Garner Allen.)

Blackbird roost in swampy deciduous thicket on the Grand Prairie, Slovac, Prairie County, Arkansas, February 1951. By systematic sampling in the roost and along roost flight lines, Fish and Wildlife Service biologists E. R. Kalmbach, Johnson A. Neff, and the author estimated the roosting population in this fourteen-acre tract at 20 million birds.

THE SLOVAC THICKET

F OR ITS SIZE, the fourteen-acre Slovac Thicket, located in the heart of the Grand Prairie near Stuttgart, Arkansas, packed the most wildlife excitement per acre that I have ever known. Each winter during the early 1950s, 20 million blackbirds roosted in the Slovac Thicket. This vibrant myriad of birds was more spectacular to me than the great sea bird colonies of Bonaventure Island in the Gulf of St. Lawrence and the egret-ibis rookeries of Florida and Louisiana.

95

The swampy thicket is so named because of its location near the village of Slovac. Farmers from the old country have been growing rice since the late 1800s on what was formerly a natural tall-grass prairie. Rice fields now surround most of the thicket, and in winter, grain left from the fall harvest operation attracts blackbirds to the area.

A dense thicket of haw, persimmon, and honey locust forms the roosting substrate for the blackbirds. The trees average about twenty feet in height. Haw is the predominant tree, and the most suitable for the roosting of the blackbirds. Because of the use of this roost for several years by the great numbers of blackbirds, the guano or dung was a foot deep in some places.

Species of blackbirds comprising the roosting population were the red-winged blackbird, common grackle, brown-headed cowbird, and rusty blackbird. The starling, a non-blackbird species, was a roosting associate. Brewer's blackbird, a western species that also wintered on the Grand Prairie, did not roost in the thicket with the other blackbirds, but out on the prairie in rice stubble or broomsedge.

The 20 million birds of the Slovac Thicket are a segment of an estimated 200 million blackbirds that winter in Arkansas, Louisiana, Mississippi, and Tennessee sections of the Lower Mississippi Valley. Banding studies show that many of these

Branches of haw broken from weight of roosting blackbirds.
Slovac, Prairie County, Arkansas, February 1951.

Woodcock and chicks photographed by Peter J. Van Huizen. The "timberdoodle," as it is sometimes called, is essentially a bird of the shrub swamps. It likes to probe for worms in the ooze of alder thickets; and in Louisiana, its major wintering ground, it often flies out from swamps at night to forage in the pasturelands.

birds come from the Prairie Provinces of Canada and the North-Central States. In the late fall when they migrate southward to the rice belt, they join forces with the resident blackbirds.

As I observed them, blackbirds moved out of the Slovac Thicket roost each morning at about dawn and returned in the evening, usually before sunset. Most of the blackbirds preferred to disperse before settling down to feed, many miles from the roost, although the same kind and abundance of food was available in nearby fields. Some Slovac Thicket blackbirds were followed for thirty-five miles from the roost. In Texas, I followed blackbirds for forty-six and fifty-two miles, respectively, from two coastal marsh roosts to their feeding ground in the rice belt.

During the winter the great wheeling flocks glean food from the stubble fields and bottomlands, consuming among other things many tons of weed seeds. Field observations and stomach examinations revealed some interesting food choices.

Cocklebur, a food that seemed unusual, was found in gizzards of several red-winged blackbirds. The manner of extracting the seed from the seemingly tough, prickly hull was observed on several occasions. A bird would pick up a bur from the ground, fly to the limb of a tree along the border of a field, place the bur on the limb between its feet, and hack at the husk until the seed was exposed.

I have spent over a hundred nights in the huge Slovac Thicket roost gathering data. The blackbirds covered all available perches from the tops of the trees to the ground. Some red-winged blackbirds were even roosting on the ground with their feet, tail, and abdomen in the water, which was several inches deep in parts of the mucky substrate.

Cold and stormy winter nights were the best for working in the roost as the birds held tighter to their perches. On such nights I could walk about plucking birds from their perches and attaching numbered bands to their legs. Usually, as I entered the roost with a head lamp turned on, birds attracted by the light would land all over me. I would often band 300 birds in a night. On one February night in a roost located about ten miles from the Slovac Thicket, my wife and I banded 925 male red-winged blackbirds between 7:00 p.m. and 1:00 a.m. Moving around in such roosts is like walking through a snow storm!

During the late spring and summer the Slovac Thicket is the nesting locale for the little Traill's flycatcher. About twenty pairs nest in the fourteen-acre thicket. This is of special interest to ornithologists since the Arkansas Grand Prairie is the type locality—the place where Audubon discovered this diminutive flycatcher. The Grand Prairie also is the southern limit of the breeding range of Traill's flycatcher in the Lower Mississippi Valley.

A ninety-acre tract of original tall-grass prairie, the last sizeable tract on the Grand Prairie, lies next to the Slovac Thicket. An effort is being made by local conservation groups to preserve these two historic and biologically important natural sites.

THE GREAT PECAN FOREST
OF THE ARKANSAS

T HE Arkansas River has its source in the Rocky Mountains. It flows eastward across the Great Plains into the state of Arkansas, entering the Mississippi about midway between Memphis and Vicksburg. Along the last thirty-five miles of its course, the river is bordered by a rich lowland forest of mixed hardwoods and cypress sloughs. John A. Putnam, former U.S. Forest Service authority on Southern bottomland forests, refers to this type as a bluff forest or riverfront hardwoods. Standing at a higher elevation than most bottomland hardwood forests in this area, it becomes flooded only about once in ten years. It is a transition forest between cottonwood and the sweet gum-water oaks type. Except for Audubon's brief description of the area based on a trip to the territorial capital at Arkansas Post in 1820, there apparently is little information in the literature about this tract of wilderness.

I first visited Arkansas Post in 1950. At that time there were only three families where once 3,000 persons lived. Also, the river which formerly flowed by the Post had changed its course and was now some distance away.

In the twenty years that I travelled the bottomland through the Arkansas Post country, the feature of the area that impressed me most was the wild sweet pecan forest. In 1968, I still found a number of sweet pecans that were over 100 feet in height and some with a diameter at breast height of over five feet, the largest being 5.6 feet. Probably the largest sweet pecan is one at Mer Rouge, Louisiana, that is 160 feet in height.

The tract along the Arkansas with the finest pecan timber lies along a bottomland ridge in the batture, the land between the river levee and the river, a mile or so down river from where Bayou Meto flows into the Arkansas. A unique feature of the Great Pecan Forest of the ridge bottom was the virtual absence of oak trees. This I never saw elsewhere in an extensive tract of bottomland hardwoods in the Mississippi Delta. Along a half-mile transect through this tract on September 7, 1968, I noted but a single oak tree—a water oak. On numerous occasions I have taken friends into this tract and asked them to show me an oak, and they have been amazed by the absence of these trees. The almost total absence of ferns was another interesting feature of the pecan forest. The only species was a tree fern, the resurrection fern or tree polypody.

The major components of the forest were sugarberry (40%), sweet pecan (30%), box elder (15%), sweet gum (10%), green ash (3%), and American elm (2%). Other trees were mulberry, redbud, and cottonwood; the latter grew along the edge of the forest on the river side. As the tallest tree with the largest crown, the pecan was dominant. Giant cane formed the understory in part of the forest. Understory plants away from the canebrake were mainly swamp privet and deciduous holly. Vines were mostly grape, peppervine, Virginia creeper, and a greenbrier (*Smilax bona-nox*). Much of the forest understory was open, presenting a glade-like aspect.

I have never seen a better area for wild turkeys. They feed all over the area but seem partial to the park-like areas and the margins of the canebrakes. By feeding close to the canebrakes the turkeys can retreat quickly into heavy cover if disturbed or frightened.

In this forest the pecan nut takes the place of the acorn, prime food of the wild turkey over most of its U.S. range. The nut of the wild sweet pecan is about one-half to two-thirds the size of most cultivated varieties of pecans. I have been unable to detect any difference in taste.

Sugarberry or hackberry, an important component of the forest, is another much-used food of the turkey and several resident songbirds. One January day,

The author and twenty-one-pound wild turkey gobbler from the lower Arkansas River bottoms, near Pendleton Ferry, Arkansas, April 1953. This specimen is now in the U.S. National Museum. (Photo by Anna G. Meanley.)

Wild sweet pecan, eighty feet in height, Arkansas River bottoms near Pendleton Ferry, Arkansas, 1970. Black dot near top center of tree is nest of Mississippi kite. In this area kites usually select the tallest trees in which to place their nest, and in the top-most crotch that will support it.

I saw an estimated 5,000 cedar waxwings feeding on sugarberries; on another occasion, approximately 200 bluebirds and over 1,000 robins were moving through the forest feeding on sugarberries and the fruits of poison ivy, supplejack, and deciduous holly.

Contents of wild turkey crop from the Arkansas River bottomlands, southeastern Arkansas, April 1952. Food items include jack-in-the-pulpit leaves (upper left), poison ivy fruit and seeds (lower left), snails, sweet pecan nuts, and scarab beetles (center), grit and seeds of sugarberry and rattanvine (right).

In a forty-acre canebrake along the lower Arkansas, I had many interesting experiences in the early 1950s. One day as I approached the canebrake I met a group of hunters who had just killed two bobcats. I asked for and later received the stomachs of these two cats, and found in them the remains of swamp rabbits. On another occasion when walking along the edge of the canebrake I found two freshly killed wild turkeys, about fifty yards apart. Both birds had been killed the same way. The ground cover was torn up for several yards around the dead turkeys, indicating a terrific struggle. Near one of the turkeys just inside of the canebrake was a fresh scat or dropping of a bobcat, later identified by a professional mammalogist. The scat was full of turkey bones, the identification of which was confirmed upon comparison with reference material at the U.S. National Museum. The craws of both turkeys were intact, and both were packed with buttercup flowers, which formed much of the ground cover next to the canebrake where I found the turkey carcasses.

On a cold day one February, I was walking down the Arkansas River levee and noticed some large dark-looking birds perched sixty to seventy feet above the ground in some tall wild pecan trees that formed the overstory of the forty-acre canebrake referred to above. At first glance I thought that these birds were black vultures, as this species frequently was seen in this area and the canebrake is a favorite nesting place. There were about twenty birds and only one or two to a tree. Then one of the birds flew, followed by another, and another. The rapid beat of the wings was not unlike that of a black vulture, but when I saw those long, outstretched necks, I realized that the birds were wild turkeys. They did not fly to the ground but to other trees, and they remained at the same height. Now, I have occasionally seen wild turkeys perched high in tall trees at night when they were roosting, but never in the middle of the day. At first I thought that they were "budding," but wild turkeys that I have seen eating buds in trees were never near the tops of tall trees. After watching them for a few minutes I could see that they were not eating buds but were merely perched—some crouched as when roosting at night—and waiting to move back to the ground. Since wild turkeys usually do not perch in trees at sixty-to-seventy-foot heights during the day, I surmised that they had been frightened, probably by a bobcat, a common mammal in this area. When startled by other creatures, including man, turkeys can almost always escape by running, but their chief natural enemy the bobcat moves very quickly and quietly through the canebrakes.

Although the Arkansas canebrake was a quarter of a mile from the nearest surface water, on three occasions I encountered cottonmouth moccasins there. These reptiles are usually associated with wetland areas. One of the cottonmouths looked as though it had just swallowed a large prey, and upon dissection it was found to contain a copperhead. Copperheads and canebrake rattlers were also seen in the canebrake at various times.

Two Southern specialties, the black vulture and the Swainson's warbler, nested in this canebrake. So dense was the part of the canebrake in which the vulture nested that the bird simply laid its eggs on the ground instead of in the usual location beneath a fallen log or in a hollow stump.

The first Mississippi kites that I ever saw were soaring just above the tree tops of the Great Pecan Forest. They are typical birds of the riverfront hardwoods belt in this area, and are seldom seen far from the river. They nest in the tops of the tallest cottonwoods and pecans, often placing their nests at heights of 100 or more feet.

The kite, the tall pecan, and the cottonwood are all a part of the passing scene as the delta hardwoods are being displaced by soybeans.

THE EVERGLADES

The Everglades, famous for its unusual birds, tropical plants, and Seminole Indians, occupies about a million acres of the south Florida land mass between Lake Okeechobee and the Keys. Originally the Everglades covered almost twice this area; now much of the former Glades is producing sugarcane, truck crops, and cattle, while residential developments are encroaching from the coast.

Contiguous with the Everglades and lying to the west of it is the Big Cypress Swamp. As one drives along the Tamiami Trail from Miami to Naples the transition from marsh to swamp is readily discernible. The Everglades is essentially a marsh, with pockets of swamp and hammocks.

Since many dictionaries imply that the terms *swamp* and *marsh* are synonymous, it seems appropriate to point out here the essential differences: swamps are wooded, or forested wetlands; marshes are wetlands covered by grasses, sedges, and other aquatic herbaceous vegetation. Biologists and ecologists, as well as most hunters and trappers, distinguish between these two quite different physiographic communities.

When one speaks of the Everglades one mostly has in mind the vast sawgrass savannas whose bounds seem almost limitless. Sawgrass, which is not a grass but a sedge, forms extremely dense growths, and its long slender saw-edged leaves can inflict severe cuts.

The monotony of the sawgrass savannas is broken by the hammocks, bay heads, cypress heads, and cabbage palm islets.

Cypress heads and bay heads are isolated groves of cypress and bay trees in the open glades. The taller trees in the center are surrounded by smaller trees, giving the grove a hump-like appearance.

Hammock, from the Spanish word *hamaca,* a garden place, is an island of trees, shrubs, vines, ferns, and air plants, that is slightly higher and thus drier than the surrounding sawgrass glades.

Most large hammocks—such as Mahogany Hammock in Everglades National Park—are surrounded by a shallow moat. This is formed by organic acids that wash out from decaying vegetation in the hammock. The acid dissolves the lime rock and the soil around the hammock edge. The moat, filled with water, helps protect the hammock from fire. Hammock plants growing outside the moat are burned off, restricting the size of the hammock.

The hammocks of the southern Everglades are a sort of meeting place of tropical and temperate zone flora and fauna. This phenomenon is well illustrated by Robertson, who states that, "In the course of a walk through a hammock in the Everglades National Park, one might observe, say, a raccoon, a blue jay and a black snake—nothing tropical about any of these creatures; they could be seen

The Everglades is a marsh with tree islands, ponds, and sloughs.

as readily in Illinois or Connecticut. However, the tree that the raccoon climbs, the insect the blue jay catches and the lizard that the black snake has just eaten are likely to be the tropical species not found elsewhere in the United States."[1]

Royal Palm Hammock and Mahogany Hammock in Everglades National Park are tropical hardwood hammocks mostly of West Indian plants, and are a botanist's paradise. Seeds of these plants, originating in the West Indies, drifted northward in the Gulf Stream or were wind-borne by hurricanes to the southern Everglades.

Dr. John K. Small, eminent authority on Southern plants, recorded 162 species of native flowering plants in Royal Palm Hammock. Seventy-five species of woody plants are known.

Gumbo-limbo, lancewood, poisonwood, mahogany, royal palm, and strangler fig are but a few of the common West Indian trees of Royal Palm Hammock.

Royal Palm Hammock in Everglades National Park is composed of many West Indian trees, including gumbo-limbo, mahogany, royal palm, poisonwood, lancewood, and strangler fig.

The life history of the strangler fig is indeed unique. A seed of this species may be deposited by a bird on an upper limb of a tree, often a live oak or mahogany. As the root system of the fig develops, it gradually entwines the host. And as the roots tighten, the food-carrying layer of the oak or mahogany will be pinched off and the tree will die. The strangler fig then becomes a tree in its place.

The branches and trunks of many of the trees of the hammocks and other tree islands of the southern Glades support aerial gardens of orchids and bromeliads or air plants. Some of the bromeliads resemble pineapple tops, and indeed belong to the pineapple family. Such plants are not parasites, and derive their nourishment from the air. Rainwater collects at the base of the leaves, and insects and tree frogs make this their home.

Many species of ferns and vines contribute to the lushness of these tropical jungles. Royal Palm Hammock has eighteen species of ferns, one of which, the leather fern, is taller than a man. Medicine vine is the most common woody vine, and is so named because of its use in Caribbean folk medicine.

The bird life of Royal Palm Hammock would seem to be inconsistent with the vegetation. Instead of finding parrots or macaws in this tropical jungle, one mainly sees catbirds, cardinals, and other temperate zone species. The same can be said of most of the other vertebrate animals. Raccoons, bobcats, and diamondback rattlesnakes occur in these West Indian-type hammocks.

The invertebrate animals, however, include many associated with the West Indies. Of special interest, partly because they are rather conspicuous, are the tree snails. They appear to be found mainly in the hammocks. Some of these are two to three inches in size, and quite colorful. Solid color forms of orange, black, and white are noted; while others are striped with various color combinations.

But my choice as the most fascinating creature of these West Indian hammocks is the zebra butterfly, a tropical species. They are common at Royal Palm and Mahogany Hammock, and in suitable habitats elsewhere in the Glades. The zebra

Strangler fig doing its work on a live oak in Mahogany Hammock, Everglades National Park. This began when a bird ate a fig and deposited a seed on an upper branch of the live oak. As the root system of the fig developed, it entwined the host. As the roots tighten, the food-carrying layer of the oak will be pinched off and the tree will die. The strangler fig then becomes a tree in its place.

is a medium-sized butterfly, long and narrow laterally, with yellow stripes across a black background. The flight of this beautiful creature is so soft and effortless that it appears to float through the hammock jungle. It is the tamest butterfly that I have seen.

I saw a dozen or so in Royal Palm and also in Mahogany Hammock. In the evening they gather in groups, sometimes numbering twenty-five or more, to roost in a favorite spot in the hammock. The whole roosting group may sleep on a single branch, often returning to the same branch or tree night after night.

The greatest spectacle of the Everglades is the large wading birds—the herons, egrets, ibis, and spoonbills. Whether in the nesting rookeries south of the Glades in the coastal mangroves or circling against the blue sky over the sawgrass, this is what most of the visitors come to see. Anhingas or water turkeys sometimes nest with this assortment of birds, and the great man-o-war bird or magnificent frigatebird comes in from tropical seas to rest in the rookeries. The roseate spoonbill, being the rarest and most colorful and conspicuous, arouses the most

The coastal mangrove forest borders the southern Everglades. Most of the large nesting rookeries of heron, egret, and ibis are located in the mangroves. (Photograph courtesy National Park Service.)

interest among the many long-legged, long-billed birds. The adults are pink and have an odd spatula- or spoon-shaped bill. The wood ibis is usually the next large wader on the birdwatcher's list of desiderata. It is the only North American representative of the stork family, and is becoming better known as the wood stork. In late summer many wood ibis wander north into the Lower Mississippi Valley where they frequent the oxbow lakes and ponds of the Delta. In Arkansas, where I have also seen them in the ricefields, they are called gourdheads and flintheads. Virtually all of the wood storks in the country nest in south Florida; however, a small colony was recently found nesting in Okefenokee Swamp, Georgia.

Among the more than 300 species of birds that inhabit the Everglades and south Florida country, there are four that are extremely rare—the Everglade kite, the short-tailed hawk, the Audubon's caracara, and the Cape Sable seaside sparrow. The Cape Sable sparrow (as it is also known) was the last North American bird to be described. Arthur H. Howell discovered this latest addition to our avifauna

Wood ibis or wood stork. The only true North American stork at nesting colony in mangroves. Most of the wood storks in this country nest in Everglades National Park and Corkscrew Swamp in south Florida. (Photograph by Luther Goldman.)

in 1918 in the marshes at Cape Sable. So surprised and elated was Howell with his discovery, that he gave the bird the specific name *mirabilis,* meaning miracle (the complete technical name is *Ammospiza mirabilis*).

The 1968 edition of the U.S. Fish and Wildlife Service's *Red Book of Rare and Endangered Species* estimates the existing population of Cape Sable sparrows at 1,000 birds. The population shifts about locally, but recently there have been several loose colonies in the Ochopee and Everglades City area in the southwestern Everglades. The *Red Book* makes the following statement relative to its rare status: "Endangered because very rare and constantly changing population due to unstable habitat result [*sic*] from drought, fires, hurricanes, encroachment of mangroves on marsh grass, and reduction of habitat by real estate development." It seems truly a miracle that this little ground-inhabiting sparrow still exists when one

Below left: The limpkin, like the Everglade kite, is another Florida specialty that feeds principally on the apple snail (*Pomacea paludosa*). The limpkin occurs in freshwater marshes and swamps. (Photograph by Rex Gary Schmidt.) *Below right:* The roseate spoonbill is one of the rare birds of the Everglades. It is known by its spatula-like bill and pink color. (Photograph courtesy National Park Service.)

In the United States, the Everglade kite is found only in southern Florida. By latest count (1969) there were only 120 of these rare birds. (Photograph by Luther Goldman.)

considers the six-foot tidal waves and 100-mile-an-hour winds that sometimes sweep over its habitat during a hurricane.

The best known of the Everglades rare birds is the Everglade kite, a raptor, or member of the hawk family. Birdwatchers from all over the country come here to add the "snail hawk" to their life-list of birds. In 1970, Paul Sykes, a Fish and Wildlife Service biologist who is studying the ecology of the Everglade kite, estimated the population at 120 birds. This is the entire North American population of this species. However, several subspecies or geographic races, differing slightly from our subspecies, occur in Cuba, Mexico, and Central America. These subspecies may also be facing the same dilemma as ours, through destruction of habitat and wanton shooting.

The Everglade kite has very specialized food habits for which it has become adapted through the structure of its sharply curved and pointed beak. It is virtually a one-food bird, its food being the apple snail (*Pomacea paludosa*). This snail, about the size of a golf ball, is fairly widely distributed in suitable marsh and swamp habitat of peninsular Florida. But the Everglade kite, whose range once corresponded with the snail's, is now found mostly in the Lake Okeechobee and eastern Everglades region of southern Florida.

In hunting for apple snails, the Everglade kite covers a marsh until it sees its prey, then it drops down to snatch the snail from the surface of the substrate in one jerk without having to land. It repairs to a favorite perch to remove the snail from its shell. The snail usually is ingested whole.

The limpkin, a large rail-like bird and another Florida specialty, also feeds mostly on apple snails. Not yet a rare or endangered species like the Everglade kite, it may be able to hold on as it is more sedentary than the "snail hawk," and a great many limpkins have discovered the good life of Florida's many wildlife sanctuaries. Most limpkins occur in the Everglades, but a few are sparsely distributed through the state as far north as Wakulla Springs, near Tallahassee; and occasionally one wanders across the border into Okefenokee.

Audubon's caracara, a buzzard-like bird, is another south Florida specialty formerly common in that area, which is now nearly extirpated in the state. It used to occur in the Kissimmee Prairie region just north of Lake Okeechobee, and in the northern Everglades. The big lake is the center of the fast disappearing Florida population. Surveys made in 1971 revealed none of these birds. However it is possible that a few may have been overlooked. The only other places in North America where this species occurs are southern Texas and southern Arizona. The caracara feeds mainly on carrion, reptiles, and amphibians. They are slow-flying and relatively tame birds, and this has led to their demise, as such large birds are prime targets for indiscriminate gunners.

No bird is so striking and graceful in the Everglade skies as the swallow-tailed kite. The breeding range of the swallowtail formerly extended well up the Mississippi Valley, and included most of the lower or Coastal Plain South. And now, except for a scattering of a few nesting pairs along the South Carolina and Georgia coast, most of the remaining population is in the Everglades and surrounding south Florida country. Arthur H. Howell briefly describes its haunts and habits in his book, *Florida Bird Life*: "The Swallow-tail Kite seems to prefer a semiprairie country, or a region of open pine glades dotted with small cypress swamps. In such sections the birds hunt a good deal over the fresh-water marshes and wet glades, flying at such times rather near the ground in their search for food. At other times however, they ascend to great heights and perform swift and skillful evolutions in a most graceful manner. They are fond of soaring over lakes or streams, into which they frequently dip to drink while flying. They are on the wing the greater part of the day and often devour their prey as they sail leisurely along. Their ordinary note is a shrill high-pitched whistle or squeal of three syllables." The food of the swallow-tailed kite consists mainly of reptiles, amphibians, and insects.

In the Everglades National Park, according to Robertson, the swallowtail arrives in late February or March, and migrates south again in September. They are,

A young mangrove in the coastal brackish water zone of Everglades National Park. The stilt or prop roots are characteristic of this plant.

he says, "among the least earthbound of birds. . . . In midsummer, after their nesting is completed, swallow-tailed kites often assemble in pre-migratory gatherings of as many as several hundred birds. The effortless aerobatics displayed on such occasions constitute one of the great sights of the bird world."

Next to the Cape Sable sparrow, I suppose that the least known bird of the Everglades is the short-tailed hawk. The short-tailed is a buteo or soaring type of hawk, like the red-tailed and red-shouldered hawks. Although it is a big bird and not too difficult to identify, it is rarely seen, as the total population is estimated at only 200 individuals in the U.S. Fish and Wildlife Service's *Rare and Endangered Wildlife of the United States* (1968).

Among the rarer mammals of the Everglades certainly the Florida panther or cougar is the most notable. It is estimated that only 100 to 300 individuals remain. The latter figure would seem high.

It seems inconceivable that the panther still exists in southern Florida. Survival has been possible only because of the extensive wilderness of the western part of the Everglades and the adjacent Big Cypress, along with sufficient prey species. This is the last place in eastern North America where they are definitely known

An alligator in its pool four feet below the water's surface in the Florida Everglades. The bottom of this gator hole is limestone rock, which underlies the Everglades. Such holes, opened up by an alligator's activity in the submerged vegetation and mud of a pond or slough, attract fish upon which the big reptile feeds. Note fish swimming about the alligator. Fishermen of the Southern swamps know about these gator holes.

to occur. If more protected lands are added to the present sanctuaries within their range, that is, Everglades National Park and Corkscrew Swamp, the big cat has an outside chance of survival in southern Florida.

No creature is more closely identified with the Everglades than the alligator. An authority on this ancient reptile estimated the alligator population of the Glades at one million before the days of the commercial hide hunter and water manipulators. In recent years the population may have dropped to 1 percent of this figure.[2] However, the population has responded rather favorably to protection offered by the Everglades National Park and several of the sanctuaries along the periphery of the Everglades. Otherwise the alligator would be close to extinction in all of south Florida.

It is not uncommon to see in protected areas of the Glades alligators measuring twelve or thirteen feet in length. There are authentic records of up to eighteen feet. The larger bulls weigh as much as 600 pounds. They grow at a rate of

The Florida panther or cougar. The Everglades and adjacent Big Cypress of south Florida are the only place in eastern North America where this mammal is still found in the wild. Probably no more than 100 occur in this area. (Photograph by Ernest Christenson, courtesy National Park Service.)

about one foot a year until they are about ten years old. They are long-lived, and some are known to have reached the age of thirty or more years.

Alligators build a mound-like nest in which up to about forty or fifty eggs are deposited, to be hatched by the protection and warmth of decaying vegetation. Nests that I have seen were usually four to five feet in diameter and two to three feet in height; they were made from marsh vegetation and mud gathered nearby.

Because of their size and habits, alligators play an important role in the ecology of marshes and swamps in which they reside. Their water holes and trails are utilized by many organisms. They create their own fishing holes, which serve as traps in which their victims concentrate. Since gator holes usually are deeper and have more light than the surrounding bodies of water, they contain more fish, and the knowledgeable fisherman knows about this sort of bonanza. These fish and turtle "traps" have their beneficial side effects, as they are used by birds and mammals as feeding and watering places during droughts and the dry winter period in the Glades when they are often the only water holes left. Herons, egrets, ducks, grebes, rails, gallinules, coots, otters, mink, and muskrats frequent gator holes and occasionally become victims of the big saurians. However, the main foods of alligators in most places are fish, turtles, crustaceans, and frogs.

A most interesting and informative account of the ecology of the alligator has been written by Frank Craighead, Sr. It appeared in *The Florida Naturalist* in 1968, under the title of "The Role of the Alligator in Shaping Plant Communities and Maintaining Wildlife in the Southern Everglades." *The Florida Naturalist*, published by the Florida Audubon Society, Maitland, Florida, is one of the best sources of information on the Everglades. Two exceptionally fine books on this area are *Everglades—The Park Story* by William B. Robertson, Jr., and *Everglades: River of Grass* by Marjory Stoneman Douglas.

The best way to see the Everglades is to go to Everglades National Park. The park headquarters is located a few miles southwest of Homestead, Florida. A hardtop road leads from park headquarters to Flamingo at the southern tip of the Florida Peninsula, on Florida Bay. Along this thirty-eight mile route are hammocks, sloughs, prairies, and ponds where rare and unusual birds and plants, alligators, other reptiles and amphibians, and a galaxy of butterflies may be observed.

THE BIG CYPRESS

The Big Cypress is the name given to an extensive area of cypress strands that are located in deep southwest Florida's Collier County. The "big" refers to the size of the area, rather than the size of the trees, although there are some huge cypress trees in Corkscrew Swamp, and formerly in Fakahatchee and other strands.

Most of the Big Cypress is made up of long finger-like strands, with wet prairies and pinewoods in between.

Corkscrew Swamp is the only remaining strand with a sizeable forest of virgin cypress, the largest such tract in North America. It is located near the Gulf Coast, about twenty-five miles southeast of Ft. Myers. This attractive area of over 10,000 acres is maintained by the National Audubon Society as a plant and wildlife sanctuary. The Society acquired Corkscrew primarily for the purpose of preserving the stand of virgin cypress, and one of the few nesting grounds of the wood ibis or wood stork, a species that probably would cease to exist without the help that it has received from the Audubon Society and the National Park Service.

The Audubon Society has built a boardwalk over a mile long through this cypress wilderness. Sections of the wilderness along the boardwalk are rather dense, and with the collection of big cypress trees, smaller trees and shrubs, ferns, vines, and aerial gardens of bromeliads and orchids, present a jungle-like aspect. If the giant cypress trees were missing, the area would look much like some of the West Indian hammocks of the southern Everglades. In fact, some of the same plants occur in both places—strangler fig, royal palm, leather fern, and various epiphytes or air plants.

The size and age of the big cypress trees in Corkscrew is impressive. Some of the trees are 125 feet in height, with a diameter at breast height of five to six feet. Cores taken from the largest trees indicate an age of over 700 years. Some of the giants have reached "old age" and have young replacements growing nearby that are only 200 to 300 years old.

Pond apple or custard apple is one of the most abundant and conspicuous of the smaller trees that form the understory of the cypress forest. It has large, dark-green, waxy leaves, and produces a fruit about the size of an apple, in late summer and fall. A few of these greenish-yellow fruits were still hanging on the trees when I visited Corkscrew one late October; and the seeds of the fruit were strewn all along the boardwalk where the raccoons left their "calling cards."

Ferns of many species are the dominant low vegetation of the swamp. Some grow on logs, some on trees, and others in the mucky substrate of the swamp.

A view in Corkscrew Swamp in the Big Cypress region
of southern Florida. Vegetation covering pond
in foreground is water lettuce (*Pistia stratiotes*).

Opposite, top: View through piney woods looking
toward Corkscrew Swamp.

Opposite, bottom: Boardwalk through a part of
Corkscrew Swamp. Small tree in foreground with dark
shiny leaves is pond apple.

Above: Bromeliads, a type of epiphyte or air plant, growing on trunks of cypress trees. Bromeliads belong to the pineapple family. They are not parasitic, and obtain their nourishment from the atmosphere.

Right: Yellow catopsis, one of the less common bromeliads or epiphytes, growing on trunk of cypress in Corkscrew Swamp. Water collects in the overlapping basal leaves of this epiphyte, and in dry periods, tree frogs make this their home.

Jungle-like section of Corkscrew Swamp. Note epiphytes
or air plants on limbs of pond apple trees.

One species, the leather fern, attains a height of ten feet in some places. Other
species include the resurrection, strap, swamp, royal, and wild Boston ferns.

The epiphytes or air plants, which include the bromeliads and orchids, are a
part of the aerial gardens seen along the branches of the pond apple and on the
trunks of the cypress. Epiphytes are not parasitic, obtaining their nourishment
from the atmosphere. Spanish moss is an epiphyte, belonging to the pineapple
family, Bromeliaceae.

If one visits Corkscrew in the winter he is treated to the sight of several thousand
nesting wood ibis and egrets. However, when I visited Corkscrew in late October
1971, what impressed me most about the birdlife was that it was the same that

Left: Young pileated woodpecker. *Right:* Pileated woodpecker at nest in river birch. Its nest hole usually is triangular-shaped. In the Coastal Plain, this crow-sized woodpecker is associated with the larger swamp and riverbottom forests.

I see around my home in Maryland about a month earlier. In addition to the same migrating warblers that pass through my home woodland in September, there was the familiar Carolina wren, tufted titmouse, cardinal, red-bellied woodpecker, crested flycatcher, catbird, red-shouldered hawk, barred owl, pileated woodpecker, and other common birds. The big pileated woodpecker is common here, but the ivory-bill, now virtually an extinct species, was last reliably reported from the Big Cypress in 1914 by F. H. Kennard, who collected one. However, the big woodpecker probably existed in Big Cypress well beyond 1914. The going is rough there, the area is extensive, and the ivory-bill's call is not so loud as the pileated's.

It is interesting that in the Big Cypress, the ivory-bill was known to forage about the cabbage palms and live oaks on the borders of the swamp. Such foraging areas exist today as they did then; yet it was thought that the birds disappeared with the cutting of the virgin or over-mature cypress and hardwoods, the reported classic foraging type for the ivory-bill.

Alexander Sprunt, Jr., reported 4,700 nesting pairs of wood ibis or wood storks in Corkscrew in the winter of 1959–60.[1] Corkscrew is an ancestral breeding ground of the wood ibis; and as early as 1912, the Audubon Society posted a warden there to prevent devastation of the nesting colony.

Apparently the nesting period of the wood ibis, which is usually in winter, is timed with the period of falling water level, which concentrates the supply of food. Palmer reports the food of the wood ibis to be mainly minnows, but also crustaceans, mollusks, reptiles, tadpoles, frogs, small mammals, insects, and occasionally seeds. They have been known to take small alligators. Palmer also reports that "Both parents fetch water to preflight young on hot days. Parent holds bill above heads of young and 'drools' water over them; most of it runs off, but they swallow some."[2]

Apparently the wood ibis was not hunted for its plumes, as were the egrets. According to Robert P. Allen,[3] the Florida population (which is virtually the entire U.S. population) at one time probably exceeded 150,000 birds. Several large breeding colonies contained 15,000 to 20,000 or more pairs. Population decline has resulted from a gradual loss of extensive unspoiled feeding areas. This decline has been over 90 percent since about 1939. The future welfare depends upon preservation of adequate feeding and breeding grounds. In addition to the natural feeding grounds in the swamp, the Audubon Society is managing additional feeding areas nearby for this remnant population.

From the above discussion, it should not be inferred that Corkscrew Swamp constitutes most of what is left of the Big Cypress region. Extensive areas, numbering in the thousands of acres, of smaller or younger cypress trees occur between Immokalee and Everglades City to the south. That the Big Cypress still qualifies as an extensive wilderness is exemplified by the presence there and in the adjacent Everglades of the only panther or cougar population in eastern North America.

OTHER INTERESTING PLACES

During the 1960s I was introduced to some of the swamps along the lower Savannah River by local naturalists Marie and Mel Mellinger and Ivan Tomkins. One of the areas of special interest was Monkey John Swamp, located seven miles north of the city of Savannah. The swamp was named for Monkey John, a hunched-back Negro who made his living from the shingles he fashioned from its cypress trees.

The hard-to-find Swainson's warbler, the swallow-tailed kite, extremely rare in the lower Savannah River area, and the spruce or Walter's pine are to be found in Monkey John.

The spruce pine is a tree of the hammock lands of the deep Southeast, and is the only pine that thrives and propagates under dense hardwood cover. Also unlike other pines, this species does not form extensive pure stands, but grows singly or in small groups. It is easily distinguished from other pines by its smooth bark resembling that of a hardwood tree, and by its darker green foliage.

Another tree of the river swamps and river edges that I wanted to get to know soon after I began working the southern lowlands, was the Ogeechee gum or Ogeechee lime, as it is sometimes called. It belongs with the black gum and tupelo gum group (*Nyssa*), not the sweet gum. Its fruits (or drupes, as the botanist might refer to them) are edible and are put up as preserves in Savannah and some of the other towns within the restricted range of this species.

Ogeechee gum is found only in the southeastern corner of South Carolina, the lower Coastal Plain of Georgia, and in northern Florida. It grows quite profusely in the swamps along the lower Ogeechee River, the next river down the coast from Savannah. It is a characteristic tree along several of the rivers that cross the line from Georgia to Florida, such as the Suwannee, Alapaha, and Withlacoochie. The St. Mary's River, which has its source in Okefenokee Swamp, and flows eastward out of the great swamp as part of the boundary between Georgia and Florida, is lined with these trees along much of its course.

When I visited the Ogeechee River with Ivan Tomkins on May 11, 1964, the Ogeechee gum was in flower; and when I crossed the Ogeechee on October 27, 1965, its attractive reddish fruits were ripe. The fruit is 1 to 1½ inches long.

In 1938, E. A. McIlhenny published an article in the *Auk*, a scientific ornithological journal, concerning a local Florida sandhill crane population in coastal Mississippi. Apparently this was the first report in a scientific journal of this little known population.

125

Opposite: Hammock-like forest in Monkey John Swamp, South Carolina, about seven miles north of Savannah, Georgia. Tree in center is spruce pine. Unlike other pines this species does not form solid stands. From a distance its bark appears like that of a hardwood tree. Tree on right is water oak, on left sweet gum. U.S. Department of the Interior biologists in picture are John S. Webb and E. O. Mellinger.

According to a report by Valentine and Noble in 1971 the Mississippi colony is a remnant of a population that once extended from western Louisiana to Georgia and peninsular Florida. In 1969 they estimated that there were only thirty-eight to forty birds left. Today, breeding colonies of the Florida sandhill cranes are found only in southern Mississippi's Jackson County, Okefenokee Swamp in Georgia, and Florida (mainly the southern portion).

Several ornithologists think that the Mississippi population may be morphologically distinct from the Florida form that occurs in Okefenokee and

Ogeechee gum in backwater swamp along Withlacoochie River, about ten miles north of the Florida line, Lowndes County, Georgia, December 1967.

Sandhill cranes gliding in for a landing. In the South, resident populations of this species are found in Okefenokee, coastal Mississippi, and peninsular Florida. There are only about forty birds remaining of the Mississippi population. (Photograph by Luther Goldman.)

in Florida; but formal taxonomic studies have yet to be made of Mississippi breeding birds. The fact that the Mississippi population occupies a habitat distinct from other cranes suggests that it may be a separate geographic form or race. Breeding cranes in south Florida and in Okefenokee occur mainly in marsh habitats, while those in southern Mississippi nest in open piney woods savannas and small open swamps ("open" meaning with a scattering of trees rather than a dense stand). Pond cypress and slash pine are the principal indicator plants of Mississippi crane habitats.

In the mid-1950s I lived in the central Louisiana city of Alexandria, which borders on Bayou Boeuf Swamp. The striking feature of this swamp is the presence of so many kinds of "horrendous creatures." Five poisonous snakes were native to the area, including the canebrake rattler, pigmy rattler, cottonmouth moccasin, copperhead, and coral snake. The last named was more closely associated with the adjacent piney woods, but I saw three of these rather docile snakes in Bayou Boeuf.

The pigmy rattler was new to me. In fact, the first time I saw one, I unsuspectingly picked it up. I showed it to a forester friend of mine, who quickly said I had better drop it. Its rattles were so small that I had not noticed them at first; they can be seen, however, if one looks for them.

It was surprising to find tarantulas and scorpions in Bayou Boeuf Swamp. They occur in the drier sections, but usually are more numerous in the nearby rolling piney woods. Some of the big hairy tarantula spiders were nearly three inches in length. As far as I can ascertain, the sting of neither of these two arthropods is highly toxic to man; but their appearance may be disturbing.

The armadillo was a common mammal of this central Louisiana swamp. When observed they were usually digging in the soil for grubs. I examined the stomachs of three that I picked up dead along a road that runs through the swamp, and all had been feeding entirely on beetle grubs or larvae. They do feed on a wide variety of foods including longleaf pine seed. This odd-looking mammal is extending its range northward and eastward. It is now fairly common in southern Arkansas.

Nesting colony of little blue herons, common egrets, snowy egrets, and water turkeys at Swan Lake, Jefferson County, Arkansas, June 1953.

Water turkey or anhinga, breeding bird of Southern swamps, nests in heronrys or rookeries where it sometimes expropriates active nests of herons and egrets. (Photograph by Luther Goldman.)

One of the prominent features of the Mississippi Delta is the ox-bow or old riverbed lake. An ox-bow is a crescent-shaped lake formed when a river has changed its course, leaving its curved route for a direct short cut. Some are given picturesque names. One that I frequently passed when I lived near the Arkansas River was called Moody Old River. Most ox-bows are ringed with cypress trees; some eventually become cypress or gum swamps.

One April day in 1951, when I was up on the Arkansas River levee a few miles downstream from Pine Bluff, Arkansas, I saw several water turkeys or anhingas circling over an old ox-bow lake. (The name of this ox-bow was Swan Lake—not very appropriate, as swan are extremely rare in this part of the country.) Water turkeys being near their northern breeding limit at this point in the

Young barred owls removed from their nest in a hackberry tree for photographing. The barred owl is one of four large birds that occur in all major Southern swamps. The others are the red-shouldered hawk, wood duck, and pileated woodpecker.

Mississippi Valley, I thought that it would be interesting to see if they were nesting at the lake.

This species usually is not associated with the denser part of swamps, but with the more open areas where there is a scattering of cypress, tupelo, swamp privet, willow, or buttonbush. Some of this type of vegetation covered three or four acres at one end of Swan Lake, and formed the nesting substrate for a sizeable colony of herons, egrets, and water turkeys. The composition of the heronry and approximate numbers of nesting birds were as follows: little blue heron, 200; snowy egret, 100; common egret, 70; green heron, 15; water turkey, 20.

In the Swan Lake heronry, water turkeys either appropriated occupied nests of common egrets, snowies, or little blues or constructed their own. Of twenty nests that I had under observation in 1953, at least six were originally active nests of egrets or herons. Additional nest-lining material was added to nests taken from egrets and herons.

On April 21, 1953, I saw a mated pair of water turkeys perched about two feet away from an incubating common egret. When I returned on the 23rd they had taken over the nest and the egret was standing by. The egret and its mate attempted to retake the nest when the water turkeys left to copulate several feet away; however, the male water turkey flew at the egrets and they backed off. On no occasion did I see water turkeys forcibly eject a common egret from its nest. They wait for the laying bird to leave and then move in. Also, on April 21, while I was hidden in a blind about twenty feet from a common egret's nest

I noted that when the incubating or laying bird left for a few minutes a water turkey quickly moved in and stood on the rim of the nest. In three minutes the egret returned and alighted about four feet away. It did not attempt to dislodge the water turkey, although it made threatening gestures by pointing its out-stretched neck and head with bill open, emitting guttural sounds in the direction of the water turkey. The water turkey in turn did the same. As the egret looked on, the water turkey picked up the three eggs one by one from the nest and dropped them over the side into the water.

On May 30, 1953, I observed a pair of water turkeys nest hunting. They moved from one occupied little blue heron nest to the next, forcing out the incubating or brooding herons as they made their inspection.

On the other side of the picture was the fact that, whenever the opportunity availed itself, egrets and little blue herons removed sticks from water turkey nests for use in the construction of their own. I have seen the entire nest of a water turkey destroyed during the bird's absence by egrets and herons. In many cases as soon as the young water turkeys had left their nest it was torn apart by the egrets or herons, which used the sticks in constructing or mending their own.

When egret nests are taken over by water turkeys, willow twigs are usually added and the nest becomes a much better constructed affair. Most of the egret nests are made of buttonbush twigs. When the water turkey builds its own nest

Young red-shouldered hawks cradled in the crotch of an ash tree deep in the swamp. They are fed mostly on snakes, frogs, crawfish, caterpillars, and rodents. The red-shoulder is *the* hawk of the bottomland and swamp forests.

Merrisack Lake in the Arkansas River wilderness, Arkansas County, Arkansas. Audubon came through here on the way to Arkansas Post (located two miles from the lake) in December 1820. Photographed August 1966.

at Swan Lake, willow branches with foliage are nearly always used. Such nests are usually smaller and more compact than those of the common egret.

Two of the places down stream from Arkansas Post, in the Arkansas River bottomlands, that were of special interest to me were Merrisach Lake and Nady. The south end of Merrisach almost touches Nady. Merrisach Lake was a wilderness lake where I never saw any other person except during the duck hunting season, and where I watched Mississippi kites swoop down low over the water to catch dragonflies, then rise 500 feet in the sky to feed on them. Their aerial maneuvers, soaring, pitching upwards, and diving, are the most magnificent of any bird that I have ever seen.

Nady is on the map and once had a post office, but when you get there you know you are at the end of the road. This was the jumping-off place to the lower White River-Arkansas River wilderness. And when I went down into that area I always had the feeling that I had stepped back about two centuries into the past. There was a sense of wilderness there that I never felt in the heart of Okefenokee, the Great Dismal Swamp, or the Everglades. Few naturalists had ever spent any length of time there, thus there was very little information in

Menard Mound in the Arkansas River bottoms near Nady, Arkansas. The Indians of the Southeastern and South-Central States mostly lived in or near the river bottoms. This is where the greatest abundance of game and fish was to be found.

the literature about this area. There had been a few settlers in those bottoms many years ago, but most of them had left before the Civil War.

As I wandered through the pecan forest and the cypress sloughs, I often thought it would have been interesting to have been there 500 years ago to see what the river bottomland looked like in the primeval state. However, life in that bottomland 500 years ago must have been far from easy. The grave of an Indian who lived along the White River in Arkansas some hundreds of years ago was excavated recently at the site of some government construction work in the river bottomlands. The mummified body was sent to the Bureau of Ethnology at the Smithsonian Institution for study, and the contents of the Indian's stomach, being well preserved, were sent to my colleague in the Department of the Interior, Francis Uhler, for examination. He identified the food items as acorns and sumac (*not* poison sumac) berries. Acorns were a rather common food of the Indians; but sumac berries tell us another story. Sumac berries are hardly more than indigestible stone and are but rarely taken as food by wildlife, which means that they are a food of last resort. Thus, it would seem to indicate that our Indian was eking out a pretty tough existence, and apparently was on a starvation diet!

Notes

References are to author, year of publication, and page numbers of works listed in the bibliography.

Introduction
1. Tanner 1942: 37.

Okefenokee Country
1. Harper 1927: 237.
2. Harper 1927: 250.
3. Cypert 1961: 501.
4. Hebard and Street (in Burleigh 1958: 365).
5. Wright and Harper 1913: 500.
6. Wright and Harper 1913: 497–498.
7. Cone and Hall 1970: 14.

The Lower Altamaha
1. Harper 1942: pl. 14.
2. Tomkins 1965: 294.
3. Chamberlain 1961: 459.
4. Murphey 1937: 9.

Canebrakes of the Ocmulgee
1. Harper 1958: 260.
2. Meanley 1966: 155.
3. Bent 1953: 36.
4. Brooks and Legg 1942: 82.

The White River Wilderness
1. Cypert and Webster 1948: 228.

I'On Swamp
1. Sprunt and Chamberlain 1949: 457.
2. Stevenson 1938: 36.
3. Amadon 1953: 463–465.

The Edisto and Brier Creek
1. Audubon 1834: pl. 198.
2. Audubon 1834: 564–565.
3. Audubon 1834: 564.
4. Faxon 1896: 207.
5. Brewster 1885: 66.
6. Brewster 1885: 468.
7. Perry 1886: 188.

Reelfoot
1. Steenis 1945: 7.
2. Ganier 1932: 28–29.
3. Spofford 1943: 25–27.

The Tensas
1. Tanner 1942: 32.
2. Agey and Heinzmann 1969: 43.
3. Catesby 1731 (in Bent: p. 334).

The Everglades
1. Robertson 1971: 21.
2. Craighead 1968: 4–5.

The Big Cypress
1. Sprunt 1961: 36.
2. Palmer 1962: 514.
3. *Ibid.*

Bibliography

Agey, H. Norton and George M. Heinzmann
 1969. Ivory-billed Woodpecker in Florida 1969. *Birding,* 3:43.

Aldrich, John W.
 1964. In *Song and Garden Birds of North America.* National Geographic Society, Washington, D.C.

Amadon, Dean
 1953. Migratory Birds of Relict Distribution: Some Inferences. *Auk,* 70: 461–469.

American Ornithologists' Union
 1957. *Check-list of North American Birds,* 5th edition. Baltimore.

Arthur, Stanley C.
 1937. *Audubon—An Intimate Life of the American Woodsman.* Harmanson, New Orleans.

Audubon, John James
 1834. *Birds of America,* volume 2. Robert Havell, Jr., London.

Audubon, John James
 1834. *Ornithological Biography,* volume 2. Adam and Charles Black, Edinburgh.

Bent, Arthur C.
 1953. *Life Histories of North American Wood Warblers.* U.S. National Museum Bulletin 203. Smithsonian Institution, Washington, D.C.

Bent, Arthur C.
 1939. *Life Histories of North American Woodpeckers.* U.S. National Museum Bulletin 174. Smithsonian Institution. Washington, D.C.

Brewster, William
 1885. Swainson's Warbler. *Auk,* 2: 65–80.

Brewster, William
 1885. The Nest and Eggs of Swainson's Warbler (*Helinaia swainsoni*). *Forest and Stream,* 24: 468.

Brooks, Maurice and William C. Legg
 1942. Swainson's Warbler in Nicholas County West Virginia. *Auk,* 59, no. 1: 76–86.

Bureau of Sport Fisheries and Wildlife
 1971. *Wildlife Research—Problems, Programs, Progress 1969.* Resource Publication 94. U.S. Department of the Interior, Washington, D.C.

Burleigh, Thomas D.
 1958. *Georgia Birds.* University of Oklahoma Press, Norman.

Chamberlain, B. Rhett
 1961. South Atlantic Coast Region. *Audubon Field Notes,* 15: 459.

Cone, William C. and Jewett V. Hall
 1970. Wood Ibis Found Nesting on Okefenokee Refuge. *Oriole,* 35: 14.

Craighead, F. C., Sr.
 1968. The Role of the Alligator in Shaping Plant Communities and Maintaining Wildlife in the Southern Everglades. *The Florida Naturalist,* 41 (1 and 2): pp. 3–7, 69–74, and 94.

Cypert, Eugene
 1961. The Effects of Fire in Okefenokee Swamp. *American Midland Naturalist,* 66: 485–503.

Cypert, Eugene and Burton S. Webster
    1948. Yield and Use by Wildlife of Acorns of Water and Willow Oaks. *Journal of Wildlife Management,* 12: 227–231.

Douglas, Marjory Stoneman
    1947. *The Everglades: River of Grass.* Rinehart and Company, Inc., New York.

Faxon, Walter
    1896. John Abbot's Drawings of the Birds of Georgia. *Auk,* 13: 204–215.

Ganier, Albert F.
    1932. Duck Hawks at a Reelfoot Heronry. *Migrant,* 3: 28–29.

Gersbacher, Eva O.
    1939. The Heronries at Reelfoot Lake. *Report of the Reelfoot Biological Station,* volume 111, pp. 162–180. Tennessee Academy of Sciences, Nashville.

Gleason. H. A.
    1952. *New Britton and Brown Illustrated Flora of the Northeastern United States and Adjacent Canada.* New York Botanical Garden, New York.

Gregory, Tappan
    1935. *The Black Wolf of the Tensas.* The Chicago Academy of Sciences, Chicago.

Grosvenor, Melville Bell
    1958. Corkscrew Swamp—Florida's Primeval Show Place. *National Geographic Magazine,* 113 (1): 98–113.

Harper, Francis
    1927. The Mammals of the Okefinokee Swamp Region of Georgia. *Proceedings Boston Society of Natural History,* 38: 191–396.

Harper, Francis (edited by)
    1958. *The Travels of William Bartram,* Naturalist's Edition. Yale University Press, New Haven.

Harper, Francis (annotated by)
    1942. *Diary of a Journey Through the Carolinas, Georgia, and Florida, 1765–66 (John Bartram). Travels in Georgia and Florida, 1773–74 (William Bartram).* The American Philosophical Society, Philadelphia.

Howell, A. H.
    1921. A List of the Birds of Royal Palm Hammock. *Auk,* 38, no. 2: 250–263.

Howell, Arthur H.
    1932. *Florida Bird Life.* Florida Department of Game and Fresh Water Fish. Coward-McCann, Inc., New York.

Hopkins, John M.
    no date. *Forty-five Years with the Okefenokee Swamp.* Bulletin Number 4. Georgia Society of Naturalists, Atlanta.

Kennard, F. H.
    1915. On the Trail of the Ivory-bill. *Auk,* 32: 1–14.

Kramer, P. J., W. S. Riley, and T. T. Bannister
    1952. Gas Exchange of Cypress Knees. *Ecology,* 33: 117–121.

Kurz, H. and D. Demaree
    1934. Cypress Buttresses and Knees in Relation to Water and Air. *Ecology,* 15: 36–41.

Lowery, George H., Jr.
    1943. *Check-list of the Mammals of Louisiana and Adjacent Waters.* Occasional Papers of the Museum of Zoology. Louisiana State University, Baton Rouge.

Lowery, George H., Jr.
  1955. *Louisiana Birds.* Louisiana Wildlife and Fisheries Commission. Louisiana State University Press, Baton Rouge.

Meanley, Brooke
  1954. Nesting of the Water-Turkey in Eastern Arkansas. *Wilson Bulletin,* 60: 81–88.

Meanley, Brooke
  1956. Foods of the Wild Turkey in the White River Bottomlands of Southeastern Arkansas. *Wilson Bulletin,* 68: 305–311.

Meanley, Brooke and John S. Webb
  1965. Nationwide Population Estimates of Blackbirds and Starlings. *Atlantic Naturalist,* 20: 189–191.

Meanley, Brooke
  1965. The Roosting Behavior of the Red-winged Blackbird in the Southern United States. *Wilson Bulletin,* 77: 217–228.

Meanley, Brooke
  1966. Some Observations on Habitats of the Swainson's Warbler. *The Living Bird,* Fifth Annual. Cornell Laboratory of Ornithology, Ithaca. pp. 151–165.

Meanley, Brooke
  1968. Singing Behavior of the Swainson's Warbler. *Wilson Bulletin,* 80: 72–77.

Meanley, Brooke
  1971. Natural History of the Swainson's Warbler. *North American Fauna,* Number 69. U.S. Department of the Interior, Washington, D.C.

Merriam, C. Hart
  1895. Revision of the Shrews of the American Genera Blarina and Notiosorex. *North American Fauna,* Number 10. U.S. Department of Agriculture, Division of Ornithology and Mammology, Washington, D.C.

Miller, Gerrit S., Jr. and Remington Kellogg
  1955. *List of North American Recent Mammals.* United States National Museum Bulletin 205. Smithsonian Institution, Washington, D.C.

Murphey, Eugene Edmund
  1937. *Observations on the Bird Life of the Middle Savannah Valley 1890–1937.* Contributions from the Charleston Museum IX, Charleston, South Carolina.

Palmer, Ralph S.
  1962. *Handbook of North American Birds,* volume 1. American Ornithologists' Union, New York State Museum, and Science Service. Yale University Press, New Haven.

Perry, Troup, D., Jr.
  1886. Nesting of Swainson's Warbler. *Ornithologist and Oologist,* 11: 188.

Putnam, John A.
  1951. *Management of Bottomland Hardwoods.* Southern Forest Experiment Station. Occasional Paper 116. U.S. Forest Service, New Orleans.

Robertson, William B., Jr.
  1971. *Everglades—The Park Story.* University of Miami Press, Coral Gables, Florida.

Small, John Kunkel
  1964. *Ferns of the Southeastern States* (reprint). Hafner Publishing Company, New York.

Spofford, Walter R.
  1943. Peregrines in a West Tennessee Swamp. *Migrant,* 14: 25–27.

Sprunt, Alexander, Jr.
1961. Emerald Kingdom. *Audubon Magazine,* Volume 63 (1): 25–40.

Sprunt, Alexander, Jr. and E. Burnham Chamberlain
1949. *South Carolina Bird Life.* University of South Carolina Press, Columbia.

Steenis, John H. and Clarence Cottam
1945. *A Progress Report on the Marsh and Aquatic Plant Problems: Reelfoot Lake.* Report of the Reelfoot Lake Biological Station, volume 9. Tennessee Academy of Sciences, Nashville.

Stevenson, Henry M.
1938. Bachman's Warbler in Alabama. *Wilson Bulletin,* 50: 36–41.

Stieglitz, Walter O. and Richard L. Thompson
1967. *Status and Life History of the Everglade Kite in the United States.* Special Scientific Report Wildlife No. 109. Bureau of Sport Fisheries and Wildlife, U.S. Department of the Interior. Washington, D.C.

Tanner, James T.
1942. *The Ivory-billed Woodpecker.* Research Report No. 1. National Audubon Society, New York.

Tomkins, Ivan R.
1965. Swallowtailed Kite and Snake: An Unusual Encounter. *Wilson Bulletin,* 77: 294.

Valentine, Jacob M., Jr. and Robert E. Noble
1971. A Colony of Sandhill Cranes in Mississippi. *Journal of Wildlife Management,* 34: 761–768.

Wright, Albert H. and Francis Harper
1913. The Birds of Okefenokee Swamp. *Auk,* 30: 477–505.

Wright, Albert H.
1932. *Life Histories of the Frogs of Okefinokee Swamp, Georgia.* The Macmillan Company, New York.

Wright, Albert H. and Anna A. Wright
1932. The Habitats and Composition of the Vegetation of Okefinokee Swamp. *Ecological Monographs,* 2: 110–232.

Index of Vertebrate and Invertebrate Animals Mentioned in Text

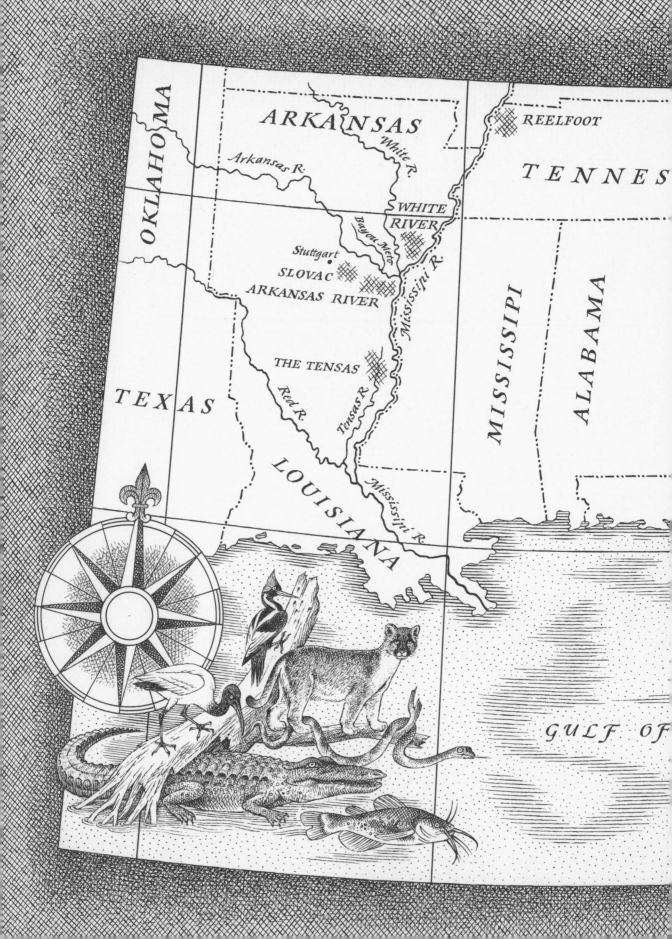

OKLAHOMA

ARKANSAS

White R.

Arkansas R.

WHITE
RIVER

REELFOOT

TENNES

Stuttgart

SLOVAC

ARKANSAS RIVER

Bayou Meto

Mississipi R.

THE TENSAS

MISSISSIPI

ALABAMA

Tensas R.

Red R.

TEXAS

LOUISIANA

Mississipi R.

GULF OF